Cognitive Theoretical Foundations of Creativity

Karl Testor

Cognitive Theoretical Foundations of Creativity

How Creativity Evolves in the Mind

Springer

Karl Testor
A-2700
I.N.F. – Institute for Neurocognitive Sciences
and Leadership (Führung)
Wiener Neustadt, Niederösterreich, Austria

ISBN 978-3-030-61735-6 ISBN 978-3-030-61736-3 (eBook)
https://doi.org/10.1007/978-3-030-61736-3

This Springer imprint is published by the registered company Springer Nature Switzerland AG
The registered company address is: Gewerbestrasse 11, 6330 Cham, Switzerland

Foreword

Psychology has been able to explain much in the comparatively short time in which it has replaced itself as an independent science of two parents who differ enormously from each other in their way of thinking and methodology. Some of the big names, such as Herman Ebbinghaus (whom we have to thank for explaining how new contents are remembered and how they remain "traces of memory," even if the contents can no longer be retrieved consciously), have to be visited under the nominal subject "Philosophy." This thought also applies to the psychologists of the Viennese School, in which not only the chair of Karl Bühler, but also that of Hubert Rohracher until the 1970s, was listed as "philosophy" (although at least with the addition, "with special consideration of psychology"). It was not until 1973, after his replacement, that the first established Ordinariate for "Psychology" was created in Vienna.

In the direction once called "psychology of the humanities" (recently also called cultural psychology), the researcher takes a look at the history of paradigms and, from that point of view, examines all explanatory models. However, he always regards this historical review as an attempt to understand the present from its historical development. This perspective prevents the present state from being seen as a final, ultimate truth and makes us aware that the word "ultimate" can only be understood in time. If you want to make a forecast, such as a future share price or the weather, you have to know the past, and you have to know that the present can only be understood if you have looked into the past. A correctly understood history of paradigms is futurology.

However, the other parent comes from a completely different scientific family, in which the guiding principle is scientific thinking, shaped by the research methods of medicine and physiology. It is considered particularly conclusive if laws derived from experience observations are also found in the biological realm and have a neurophysiological equivalent. Thus, by looking at the two completely different realities of physiology and psychology together, laws can be found which—here Fritz Wallner's constructive realism is to be followed—do not have a higher claim to truth, are not more "real," but allow more reliable predictions and make it possible to model demanding mental functions and to map their function in computer programs, some of which already work better than the brain of an average person.

These two directions of psychology are characterized by an irreconcilability that has lasted for more than a century. But is it really a "crisis of psychology," as Karl Bühler lamented in 1927? Is it really so unusual for a science to assume completely different initial assumptions without thereby falling into two irreconcilable camps? Is it not rather a "crisis of psychologists" who do not seem to have the tolerance to respect completely different approaches, different methods, and different starting assumptions? Could this not be measured in physics, which is otherwise always declared with enthusiasm the leading science in psychology?

According to the standard model of physics, the world consists of elementary particles, quarks and leptons, and string theories are based on one-dimensional objects vibrating in a high-dimensional space–time. It is not easy to find another example of how in a science such different points of view can coexist in the approach as in the consequences of the initial assumptions. But no physicist has ever written a "crisis of physics." Rather, there is a silent agreement that it is perfectly legitimate to devise very different models that can help to interpret reality.

Karl Testor asks himself a question for which one must have knowledge from both camps of our subject, psychology of consciousness and cognitive neuroscience: the question of how something new can arise in our thinking, how creativity can be understood, modelled, and specifically promoted.

However, the use of various models of modern cognitive theory, each of which can excellently fulfill a certain task such as recognition, classification, solving problems, or making decisions, does not initially lead further in this discussion and leaves completely open the question how something new can emerge. The answer we are led to in this text is the realization that these building blocks must not only be joined together in a suitable manner, but that their interplay must also be controlled by a hierarchical interconnection that allows the information to flow through the appropriate course settings. In this way, each individual component is activated at the correct time depending on the current excitation level. The decisive step by which we can break out of all previously used processes, however, is the creation of a state of significantly reduced activation, a "default mode" in which the system, also becoming visible in the architecture of the program, becomes open for new, previously unused settings of the course and new can emerge.

From a psychological experience point of view, this state is far removed from the state of consciousness with which we usually approach the accomplishment of a performance requirement: namely awake, active, concentrated, and with the full use of our strengths. For the system to run in a direction programmed for "novelty," a completely different attitude is required: entering into contemplative calm and switching off purposeful thinking. In the Asian culture, this is an age-old matter of course. A yogi would be most surprised not to realize that for the solution of many questions, not bright, sharp thinking, but rather entering into meditative rest, is an indispensable prerequisite. Cognitive theory and neuroscience confirm this ancient experience and Testor, through this synopsis, as he rightly states in the final sentence, "provides us with a positive approach to creating and preparing anew."

Psychology Department
Sigmund Freud University Vienna
Vienna, Austria

Giselher Guttmann

Preface

Daily life consists largely of routines and repetitive processes. New and creative approaches develop in these routines. Smaller or larger problems are solved, new ways of structuring a process are found, collaborations are changed. What is creativity and where does it come from? While travelling around the world and talking to people of different professions as well as those with different fields of knowledge, I noticed something new. We like to talk about the outstanding achievements and brilliant ideas of others, but do not realize that we often accomplish similar things every day! In creativity research, we correctly distinguish between ordinary and extraordinary creativity. But what makes the difference? Why do some people suddenly perceive things as extraordinarily creative? This is hard to say. The fact is that it is a matter of attributing genius to others, and that genius is only created by a special value that others attribute. Genius and extraordinary creativity are therefore social phenomena, or at least group-dynamic processes. If one excludes the attribution of value, the only question that arises is how can something new emerge? Can this be stimulated or initiated? What aspects could these be?

This work aims to find answers to these questions. With a focus on how new things emerge, various models from neuroscience, cognitive science, and creativity research have been used to find those aspects. Without anticipating the end, it can be said at this point that this has also been successful.

Focusing on creativity always leads to thinking and especially to neuronal aspects. Although cognitive sciences have long been concerned with learning, behaviour, or other areas of work, a functional path provides many models. At the same time, many explanations of neuronal processes are becoming more visible in medicine, which provide some answers for the functioning of the brain. Whether the brain is the cause of thinking or another life-giving impulse is excluded here. However, it is clear that there is a connection between the brain organ and thinking itself, as at the latest, when one considers topics such as influencing consciousness, sleep research, anesthesia, or ventures in everyday life with consciousness-altering substances. One major assumption for this work is at least that there is an interaction between matter and non-matter, from which we can elicit a functional understanding through the aforementioned scientific approaches.

That these questions could be realised in the way that is now available comes from the support of many. As in all areas of life, one needs luck for it. So, it is a special stroke of luck that my wife Manuela not only accompanies my life, but also fills and supports it with passion. Together with our family we find not only beautiful moments but also many inspirations. Thank you all very much!

Another coincidence for me with regard to this work was that a good friend gave me a wonderful recommendation. He said that he knew someone whose heart for the questions of thought, consciousness, and science would be as burning as mine: Univ.-Prof. Dr. Giselher Guttmann. Professor, I thank you not only for our beautiful, stimulating, and motivating discussions, but also for your support during my work. But I am especially grateful that we were able to follow this path of a common passion!

Next, I thank Univ.-Prof. DDr. Kurt Greiner, who "had my back" critically and supportively. Dear Kurt, I thank you especially for your reflections on the humanities, but also for our other interesting and exciting conversations.

Finally, I wish for my readers that they will feel the same sense of joy and interest in reading as I was given in writing!

Wiener Neustadt, Austria Karl Testor

Contents

List of Figures

List of Tables

Chapter 1
Introduction

Contents

Just to let an object float once. Or floating through the room yourself to get something from the other end of the room. What sounds more like dreaming would end in a shock moment for those trying, except for a very small group of people. This small group of people seriously expecting the pen or aftershave to float after letting go is neither physically nor psychologically failing. They simply expanded their understanding of the world, literally beyond the horizon. They are astronauts.

In addition to the known and expected changes of physiological nature caused by the lack of gravity, anomalies in astronauts' everyday behaviour also become apparent after their return. However, these are just as unsustainable as the physical ones. However, excesses of these experiences may last somewhat longer (cf. Unknown; Homepage ESA 2012. The vestibular system and "Space Adaptation Syndromes"). For example, astronauts reported that they were taking actions that would normally only have worked under gravity, like letting a pen float (cf. Mars1 2013, "Return to Earth").

Even though these anecdotes might be considered fairy tales in the sources, after their return astronauts still show physiological changes in their bone density, the muscular system, the cardiovascular system, and the nervous system. Changes are observed in the inner ear and the areas derived from it (cf. Nicolosi et al. 2015, p. 4, tables 1, 42), which can be traced back to the lack of gravity and which, although practised for years, temporarily restrict mobility on Earth. These persons have problems with their sense of balance for a long time (cf. Nicolosi et al. 2015, pp. 3, 42). With regard to cognitive and behavioural changes, NASA also cites minor to moderate consequences (cf. Nicolosi et al. 2015, p. 16f.). A stay away from the Earth and the framework conditions associated with it also seems to entail cognitive changes.

K. Testor, *Cognitive Theoretical Foundations of Creativity*,
https://doi.org/10.1007/978-3-030-61736-3_1

After a short time, fundamental behaviours change when reference systems change. The simple dropping of objects would normally lead one to wonder whether something is wrong with the person acting. Taking into account the longer stay in space, however, the evaluation scheme of the observer also changes. The observer experiences a broadening of his horizon, even though it may be less sustainable than that of the astronaut.

At this point, however, a rational-minded person will begin to ask himself: How real is our reality? How stable and safe is what I think I know? What are thinking and experience — on what do they depend? Maybe you even start to wonder what effects change would have on you.

Thus, the way is taken in the direction of that organ which is held responsible for thinking. At least, it seems to be a kind of biological prerequisite for this: the brain. It should be foretold here that thinking and the neuroanatomical fundamentals are closely interrelated. To what extent these cause thoughts cannot be said. Just as one cannot say why living beings live, by what energy, soul, or how one may describe this undefined distinguishing quality, that changes otherwise lifelessly understood matter into living. Nevertheless, the neurosciences and cognitive sciences try to deepen the understanding of the connections between thinking and the brain.

This work, however, wants to move beyond this idea and ask how such experiences of astronauts can be combined with the existing models. The question is asked how it is possible to integrate something completely new into the existing thinking and why this, once it is there, cannot simply lead to a change again. This question will move away from the specific idea of the Odyssey in Space to those areas that seek to capture thinking and creativity in everyday life. This work is about the fascination of thinking and the question of how something new can emerge from it. It is therefore also a work about creativity, both ordinary and extraordinary.

The structure of this work is based on that presented in a course at the Sigmund Freud Private University Vienna (SFU) (cf. Stephenson 2014).

1.1 Preface

When Konrad Paul Liessmann, in his opening speech for the Salzburg Festival Weeks 2016, quotes Schiller as saying that for him arts next to science is one of "the noblest tools of mankind, which allow him to move in the realm of perfect freedom"(cf. Liessmann 2016, p. 5), he then speaks of the fact that freedom and the collective contradict each other in an irritating way.

"Thoughts are free" is the title from a German folk song and later student song from the eighteenth and nineteenth centuries, which is often used in this context. If one now recalls experiments such as that of Benjamin Libet, in which he showed that a motor volitional action is preceded by a neuronal activation (cf. Libet et al. 1983), then one might have doubts about this thought. One of the deductions from this experiment was the lack of freedom of human independence on their neurobiological conditions. In an intensive discussion, a kind of human veto right was included, which Libet himself adopted for the model (Libet 1985). Nevertheless, as the result of temporal differences of about 150 milliseconds (ms), the question remained open as to whether and to what extent the free will of humans was given at all.

At this point, a tightrope walk in the attribution of what constitutes life—or thinking—becomes apparent, for it seems to show a dependence of humans on their biological roots that seem insurmountable. However, in all euphoria it is often forgotten that for an understanding of consciousness one would have to be primarily know and define consciousness, in this special case even by using aspects of the biological correlates. Therefore, these very general statements are only admissible in the specific world of neuroscience and only to a limited extent. Part of these limitations is given by the distinction between correlation and causality, because a spatially close occurrence of independent variables does not yet permit cause–effect derivation (cf. Rösler 2011, p. 400). This is only possible in very limited framework conditions, whereby these framework conditions are always limited to themselves and generalisation can only be carried out to a limited extent. At the same time, assumptions are made for such framework conditions, which at some part do not exist in the expected specification.

Slowly the topic of thinking begins to become complicated and at the same time the boundaries of this area become more visible. The neurobiological correlates of thinking are strongly brought into the focus of attention in the neurosciences and are currently experiencing positive support for this discipline because of the high expectations, despite enormous research and study costs. Nevertheless, they are overrated both medially and methodically. The imaging procedures make activation patterns and are the latest approaches to understand what happens in the brain, but not the visualisation of thinking per se. The resolution of methods such as functional magnetic resonance imaging (fMRI) or electroencephalogram (EEG) alone is so inaccurate that only tens of thousands of nerve cells can be measured together (cf. Rösler 2011, p. 9f.). As the brain is constantly active, "neutral" reference values must also be created (cf. Gramann and Schandry 2009, pp. 133, 163) to make changes visible. But these are not neutral.[1] The neurosciences are thus not blessed with a method that makes it possible to make thinking visible, but only to show the effects of it in special contexts.

Consequently, it becomes clear that understanding of these processes still requires interdisciplinary considerations of several disciplines. Mutual confirming can achieve higher certainty in accuracy of conclusions. The author therefore sees the neurosciences and their methods as tools for cognition or creativity research as well as psychology, sociology, or leadership sciences.

If one now refrains from the general separation of body and soul in the sense of body–soul dualism and assumes a connection between body and mind, whatever

[1] The brain is always active. Therefore, a so-called baseline does not refer to a state of non-activation, but to an activation state originated by a specially created reference-stimulus. In some cases the subject is asked to "not think" to create this reference stimulus, which is probably the most difficult thing for a person to do.

this may be, one begins to expect to find the origins in the search for thinking in the structure. It is already known from everyday experience that injuries to the brain or the consumption of certain stimulants lead to a change in the behaviour of individuals. For a fully functional way of thinking, the brain seems to be a prerequisite in this function, similar to the cardiovascular system for keeping the organism alive. From this understanding, the studies are valid, although a causality relationship cannot yet be derived from them.

We are thus still on the question of human freedom and its partial dependence on its biological realisation. Referring back to the quote mentioned at the beginning with reference to Schiller by Liessmann, two areas of society in particular are particularly free: science and art. In both areas, there are particularly outstanding personalities who seem to perform unbelievably. These people, often referred to as geniuses, are seen as blessed with flashes of inspiration following a sudden inspiration. This view is supported by introspective reports in which these moments are represented as if they were "suddenly supernatural and dreamlike" or by "the soul of the composer" (cf. Andreasen 2005, pp. 41, 43). Not depicted in this idealisation are those moments in which these persons do not succeed, as do the many empty phases of "normal" creativity. At the same time, however, this special form of creativity expresses an aspect associated with freedom. Creativity is breaking out of the norm, something abnormal, or even crazy. If something is crazy, that is, it goes crazy from its actual place, something new can emerge. The attribution of the special is a social achievement. Only the evaluation by others will highlight an achievement as something special. This is at the same time unbound by time, so that the recognition of an extraordinary achievement can may take place years later or even after the death of the creative personality.

Freedom is therefore a special aspect and basis for creating something new. It does not necessarily only have to be present in democratic societies, but there is a higher probability that freedom is given, as can be seen, for example, from the number of patent applications (cf. Friedman 2006, p. 675). Perhaps this is one reason why geniuses are envied for their inspiration. Does this give them a special glow of freedom?

In both the arts and (economic) science, idealisations entwine around those persons who are particularly creative. Because particularly creative achievements are at the same time associated with a decoupling of dependencies through expected monetary compensation, creativity often leads to an additional glorification of these conditions. Geniuses are therefore particularly exceptional people who perform services of which others would not be capable.

But what if that were not the case and the creative potential was inherent in everyone? Hardly any other being is as adaptable as humankind, who has settled down on all continents in all climatic regions of the earth and now strives to find something new, even extending into space. As living beings, humans are given a very comprehensive ability to solve problems as well as an imagination. In creativity research, part of this idea flows into the question of how this extraordinary creativity is possible. With the help of in-depth process research, it is possible to find approaches for promoting creativity and the factors that prevent it. A special aspect is stress. Stress limits cognitive performance (cf. Taverniers et al. 2013), and this limitation has a negative effect on mental flexibility, which is important for creativity (Plessow et al. 2012). In addition, studies show that early stress actions in young mice have a limiting effect on neuronal development (Mehta and Schmauss 2011).[2]

In the course of research into the prerequisites, the latest methods are available for science to try to fully understand this particular achievement by taking a look into the brain. Unfortunately, however, the same problems arise as the neurosciences had, namely, that correspondence did not automatically mean cause. On the other hand, the problems of creativity research remained, which expressed themselves in such a way that an extraordinary *Aha!* experience could not be forced. However, in the course of various experimental approaches on problem-solving competence or other areas, it was possible to show in an unfamiliar context that even average persons can be distinguished in mental flexibility. This approach, in conjunction with other tasks to be solved creatively, makes it possible to examine the ordinary form of creativity inherent in all persons. As a result of this quantification possibility, better insights were gained into the framework conditions, which in turn had an effect on the development of creative techniques.

As a logical consequence, the path of connection with the neurosciences was taken, which in combination with these test designs could serve as a basis for further insights. Progress in these areas is now so advanced that people can be examined during an improvised creative activity such as playing the piano (Limb and Braun 2008), as well as in connection with boredom and daydreaming (Gielas 2013), to come closer to the flashes of inspiration.

"You can't teach an old dog new tricks!" A proverb that communicates an understanding of the human being as a limited learning being. Is there not an unbelievable discrepancy in contemporary society between people's possibilities and the requirements and expectations on one's self? Lifelong learning is a term that actually suggests that life demands something from people that they cannot do, that is, that they are doomed to failure from the outset. Is this a reason for increasing depression and burnout? The failure to achieve the essential but not the impossible?

Here, too, thanks to scientific curiosity as well as luck and chance, science has shown a greatly reduced significance of this idea. Thus, in addition to the change in the interconnection structure through use, it is now also known (cf. Rösler 2011, p. 161f.) that in the course of the formation of new nerve cells, larger changes can be produced throughout life (cf. Roth and Strüber 2014, p. 254f). The brain is constantly changing. These changes are exactly the part that is now interesting for the question of the emergence of something new.

In this work a possibility becomes visible, which leads to understanding, why the brain, which is involved in thinking as an organ, enables this breaking out of structures as well as the emergence of new ones both in the structure and in the way of

[2]The transfer or results to humans is not valid without limitations but necessary for some topics and questions because of ethical constraints.

working. The approach is achieved by combining neuroscientific findings and models with those from cognitive science and creativity research.

Another important part of this work, although smaller in numbers of pages, concerns a scientific theoretical discussion on the general validity of scientific results. As already described at the beginning of this chapter, science is also about freedom. However, this should not be expressed in arbitrariness, which is why critical reflection should always serve as a yardstick for relativising insights and views as well as results.

Another aspect is the gap between the natural sciences and the humanities, which both see as difficult to overcome for a variety of reasons. This work takes place in a scientific context but will include approaches from the humanities in its relativisation. The author sees Friedrich Wallner's constructive realism, which inspired him and in which the staking out of the tasks is clearly expressed, as an essential reason for motivation.

An important basis for these implications is the assumption that all thinking is based on constructs, that is, not on representations taken from the environment but on representations constructed in thinking. A simple and everyday example can be offered here. If you paint two dots on a sheet of paper so that they are about 5 cm apart and then cover one eye with your hand, you will notice that if you look at one of the dots, it suddenly disappears at a certain distance from the sheet. This effect occurs because that the retina has no receptors at the site of the optic nerve exit, the blind spot. Consequently, it should not be possible for any perception to take place at that point of the visual field. As this is nevertheless possible, however, it must be an artificially generated construction: this becomes apparent in the self-experiment just described when the brain itself fills this spot. The adjacent receptors are used, and the signals of these receptors are simply connected to each other. Therefore, the dot disappears when it is projected through the lens exactly onto the blind spot and the surrounding material "hides" it.

In addition to this short example of the constructions, more detailed explanations follow later. What is important is that the author is taken with the meaningfulness of this and the constructive realism, as is reflected in parts of this work. At the same time, the understanding of the construction of reality is very interesting for creativity. For through the understanding of a constructed perception of the world, which at the same time should produce a proper relativisation of dogmas, paradigms, and doctrines, the necessity of creativity becomes even more important. This change is crucial for breaking out of structures and standards. Or, to deepen it with the words of the opening speech:

> But the realm of freedom, also and especially of aesthetic freedom, is not without pitfalls. Freedom also means freeing oneself from the spell of the community and collective thinking and reaching a real individuality. [...] The fascinating and disturbing thing about art to this day is that it can be everything that is attributed to it and yet never merges into it. (Liessmann 2016, p. 5)

This approach does not dissolve the human moment as a social being, but nevertheless the split nature of it becomes apparent. On the one hand, a person should be, and should want to be free, to individualise himself, and at the same time there is a striving for human community and commitment. It is characteristic in this context that this realisation of a discrepancy is also evident in creative people who are not only more creative but also more susceptible to mental illness (cf. Andreasen 2005, pp. 87, 95, 97). Maybe being very creative is not so fulfilling after all.

Because the question of the learnability of creativity has already arisen, this aspect is also taken into account briefly. In the course of connections with computer-based training programs, the so-called neurofeedback, which can be established via EEG derivations, it is possible to apply various training methods for stress reduction (cf. Hammond 2007), epilepsy (cf. Sterman and Egner 2006), or attention deficit syndrome (cf. Arns et al. 2009). Sustainable successes have been achieved, which can be traced back to a fundamental ability to learn such activity patterns. That a similar success for useful creativity activity patterns can be derived from it is likewise to be expected. But to what extent this is restricted by an individual limitation or by other factors remains to be seen in the future. This work will contribute to how these successes can be achieved without the technical tools.

In the case of technical aids, a special possibility should also be mentioned: transcranial magnetic stimulation. This type of stimulation induces a magnetic field via a coil attached to the head, through which alternating current flows. This induction happens without causing pain to the skull and in turn induces electrical current in the brain, which can excite neurons (cf. Wassermann 1998, p. 2). This idea is based on the assumption that by activating certain areas a change can be produced which can have an effect on behavioural changes. Experiments can show very individual and different effects in volunteers. In addition, the frequency-dependent effects are assessed differently (Maeda et al. 2000, abstract), which raises the question of the danger to the nervous system. On the one hand, animal studies show that a permanent hazard is not to be expected, that is, that there could be a kind of natural protective mechanism (Post et al. 1999, abstract). Nevertheless, at least temporary disturbances are confirmed that, for example, affect mood or cognition (cf. Wassermann 1998, pp. 5, 7). The spatial influence of this technology is much coarser than that of the aforementioned resolutions of imaging processes. A precise application is therefore also not possible in this area, whereby a still latent philosophical question can be posed: to what extent does this bring about anything at all?

In the course of this introduction it was now shown that creativity can be promoted and that there can also be restrictive aspects for this: these are connected with thinking, which is implicitly assumed. Thinking in itself is related to the brain, where the question of "how" cannot be answered, but there are interdependent interactions. These interactions can be shown experimentally, but they only represent a correspondence and thus a constructed assignment, but no causality. In addition, it is shown that a change in thinking is certainly possible with technical support, whereby the cause of the same eludes observation. Thinking and creativity are connected with the function of the brain. How this connection can appear is worked out in the following chapters. In the end, various aspects will reveal possibilities for increasing the creativity potential.

A further development, which the author would welcome, would be a rapprochement of scientific and humanistic positions, because conclusions should be drawn from this work in both areas as to which explanatory approaches are to be used for the biological "Reality," the limited existence of thought, and thus also the dissolution of this limitation through work. This last, possibly also utopian, wish is certainly the most improbable and distant, but it is precisely on this that science and creativity thrives that goals often believed to be improbable are named at least once.

1.2 Preview

The structure of this work is based on the form taught in a course at the Sigmund Freud Private University Vienna (SFU) (cf. Stephenson 2014). The first step in Chap. 2, Disciplinary Connection, is to break down those disciplines from which the author will draw essential foundations as a basis for this work.

The next step is to review the current state of research in Chap. 3. In this project, the basics of the disciplines are developed and the models necessary for the further investigation of the work are defined. Building on this chapter, Chapter 4 thus elaborates the research gap that leads to the research question presented in Chap. 5.

Chapter 6, Methodology, begins with an integration of the models necessary for the investigation and establishes a reference to the systematic approach in the investigation. A model of thinking is created as well as an own definition of thought. Starting from these, the direction of the analysis, which is applied in the following chapter, is further determined through the connection with creativity.

Chapter 7, Investigation, is the actual core of the work, in which the new model of thinking integrated from the various models is first examined in its theoretical effectiveness in relationship to the question of how this new approach can arise, and subsequently a reference is established with various experiments.

The results of this investigation are then elaborated in Chap. 8. The different aspects of the model of thinking from theoretical and experimental work are related. Thus, the basis for the discussion of the results in Chap. 9 is created, which now carries out the subdivision of the terms in an initial legitimation based on the theory of science, then considers the results in depth from the point of view of creativity research via a discussion, and finally relates them to the model of thinking and its definition of thought.

In Chap. 10, the limitation of the validity is described, which results from the investigation and the choice of disciplines in particular. A more general limitation of the binding character follows in Chap. 11, Disciplinary Reintegration, because this is a general scientific theoretical examination of the general question of the scope of scientific results. With the help of this basis, which is described as a reflection of results from the point of view of scientific theory, the actual connection back to the disciplines takes place, in which the insights useful for these disciplines from this work have been listed in parts.

Chapter 12, the résumé, condenses the results and leads to Chap. 13, Outlook, which briefly provides links to this work.

References

Books and Journals

Andreasen, N. C. (2005). *The creating brain. The neuroscience of genius*. New York/Washington, DC: Dana Press.

Arns, M., de Ridder, S., Strehl, U., Breteler, M., & Coenen, A. (2009). Efficacy of neurofeedback treatment in ADHD: the effect on inattention, impulsivity and hyperactivity: a meta-analysis. *Clinical EEG and Neuroscience, 40*(3), 180–189.

Friedman, T. L. (2006). *Die Welt ist flach. Eine kurze Geschichte des 21. Jahrhunderts*. Frankfurt: Suhrkamp Verlag.

Gielas, A. (2013). Hat Langeweile einen Sinn? *Psychology Today, 40*(10), 40–44.

Gramann, K., & Schandry, R. (2009). *Psychophysiologie – körperliche Indikatoren psychischen Geschehens*. Weinheim-Basel: Beltz Publishers.

Hammond, D. C. (2007). What is neurofeedback? *Journal of Neurotherapy, 10*(4), 25–36.

Libet, B. (1985). Unconscious cerebral initiative and the role of conscious will in voluntary action. *The Behavioral and Brain Sciences, 8*(4), 529–566.

Libet, B., Gleason, C. A., Wright, E. W., & Pearl, D. K. (1983). Time of conscious intention to act in realisation to onset of cerebral activity (readiness potential): the unconscious initiation of a freely voluntary act. *Brain, 106*, 623–642.

Maeda, F., Keenan, J., Tormos, J., Topka, H. G., & Pascual-Leone, A. (2000). Interindividual variability of the modulatory effects of repetitive transcranial magnetic stimulation cortical excitability. *Experimental Brain Research, 133*(4), 425–430.

Mehta, M., & Schmauss, C. (2011). Strain-specific cognitive deficits in adult mice exposed early life stress. *Behavioral Neuroscience, 125*(1), 29–36.

Plessow, F., Kiesel, A., & Kirschbaum, C. (2012). The stressed prefrontal cortex and goal-directed behaviour: acute psychosocial stress impairs the flexible implementation of task goals. *Experimental Brain Research, 216*(3), 397–408.

Post, A., et al. (1999). Repetitive transcranial magnetic stimulation in rats: evidence for a neuroprotective effect in vitro and in vivo. *European Journal of Neuroscience, 11*(9), 3247–3254.

Rösler, F. (2011). *Psychophysiologie der Kognition. Eine Einführung in die Kognitive Neurowissenschaft.* Heidelberg: Spectrum Academic Publishing House.

Roth, G., & Strüber, N. (2014). *Wie das Gehirn die Seele macht* (5th ed.). Stuttgart: Klett-Cotta Publisher.

Stephenson, T. (2014). *Deep impact*. Workshop on ZRD: Citation/Reception/Discourse. Vienna: SFU. 30/31 January 2014.

Sterman, M. B., & Egner, T. (2006). Foundation and practice of neurofeedback for the treatment of epilepsy. *Applied Psychophysiology and Biofeedback, 31*(1), 21–35.

Taverniers, J., Taylor, M. K., & Smeets, T. (2013). Delayed memory effects after intense stress in Special Forces candidates: exploring path processes between cortisol secretion and memory recall. *Stress, 16*(3), 311–320.

Wassermann, E. M. (1998). Risk and safety of repetitive transcranial magnetic stimulation: report and suggested guidelines from the International Workshop on the Safety of Repetitive Transcranial Magnetic Stimulation. *Electroencephalography and Clinical Neurophysiology 108*, (1998) 1–16.

Online Resources

Limb, C. J., & Braun, A. R. (2008). Neural substrates of spontaneous musical performance: an fMRI study of Jazz improvisation. *PLoS One, 3*(2), e1679.

Mars1. (2013). *Skylab: America's first space station.* http://ablogaboutnothinginparticular.com/?=1042. Accessed 1 Aug 2016.

Other

Liessmann, K. P. (2016). Kunst als Manifest. Plädoyer Salzburger Festspiele. In *Kleine Zeitung of 29 July 2016* (pp. 2–5). Graz: Kleine Zeitung GmbH & Co KG.

Nicolosi, L., Eggert, J., Riggenbach, S., Khan-Mayberry, N., Wilson, R., & Hensley, J. (2015). *NASA's effort to manage health and human performance risks for space exploration.* Washington, DC: NASA Office of Inspector General-Office of Audits. Report No. IG16-003.

Chapter 2
Disciplinary Connection

Contents

As has already been mentioned, by considering the subject of thinking and the brain's ability to let this new thing develop, a first field has already been defined in which this work will move. However, the neurosciences and cognitive sciences, as well as creativity research, are not the only areas in which possible findings for this topic are possible.

2.1 Medicine

The neuron is an important basis for understanding neuronal activity. Because this is very intensively investigated by basic anatomical research, especially in medicine, the author takes this expertise as an opportunity to confirm fundamental derivations of the neurosciences in this area.

An important step for the neurosciences is the conclusion reached by the imaging methods[1] that an increased oxygen demand is achieved in areas of increased blood flow through the more active neurons.

[1] Referring to imaging methods, the author here means especially positron emission tomography (PET) and functional magnetic resonance imaging (fMRI). These methods record changes of the

K. Testor, *Cognitive Theoretical Foundations of Creativity*,
https://doi.org/10.1007/978-3-030-61736-3_2

Important for understanding are the mitochondria, which are decisive for the oxidation processes and energy supply. It can also be determined that cells with high energy expenditure for metabolism or transport exhibit a large number of mitochondria (cf. Silbernagl and Despopoulos 2003, p. 12). Because neurons exhibit a large number of mitochondria (Kahle and Frotscher 2005, p. 22), it can also be concluded that their processes are energy intensive and that this intensity is reflected in imaging.

Another important aspect is provided by the neurotransmitters and neuropeptides, which have a major influence on the excitation or inhibition of neurons (Kahle and Frotscher 2005, pp. 26, 30; Silbernagl and Despopoulos 2003, pp. 50, 52). They are also jointly responsible for the plasticity and changeability of the brain (Rösler 2011, p. 162f.; Silbernagl and Despopoulos 2003, p. 336).

Based on the findings from anatomy and physiology, the first approaches for understanding neuronal networks can be found here, which together make visible the correlates of experience on the sensory level and thereby various basic rules.

2.2 Cognitive Psychology

"Cognitive psychology seeks to understand the basic mechanisms of human thought" (Anderson 1996a, p. 3).

In the actual question of what thinking is and how it develops, cognitive psychology describes the processes involved in the perception, processing, and storage of information, thereby attempting to find models of the brain that are as realistic as possible.

In this area, Anderson's ACT-R theory is an important starting point for the investigation, which considers intensively how elementary perceptions[2] can be processed and subsequently integrated into a processing sequence system. He extends this system in the *Integrated Theory of Mind* around an organic reference, that is, the different brain areas in the attributed specialisation, with his original ideas (cf. Anderson 1996b, passim; Anderson et al. 2004, passim).

An important finding of cognitive psychology is that certain preconditions can influence the speed of processing. These can lead thoughts in certain directions, such as priming, or lead to disturbances, such as interference. These effects further on allow derivations that are reflected in the activities of the brain. Showing this is the starting point for the neurosciences.

---

blood level in narrowly defined areas of the brain (cf. Rösler 2011, p. 10). These changes are associated with an increased oxygen demand in these areas, caused by a higher neuronal activity.

[2] For example, Anderson refers to "chunks" as being the elementary unit of declarative knowledge. Different data are stored in this element, but further details are not explained.

2.3 Neurosciences

> "'Neurons spiritualize' [...] It's about explaining how 'spirit' is created by the activity of the neurons." (Rösler 2011, p. 11)

The aim is not only to show the activities, but also to achieve an understanding for thinking. In this discussion, however, the author does not discuss body–mind dualism, as it might seem in the quote. However, he also acknowledges that these, as well as other approaches from the linking of self-organizing systems in connection with quantum theoretical approaches via matter–nonmatter interaction, have an essential role in the future cognitive sciences and neurosciences to gain further insights.

Rather, the author sees the neurosciences as an important additional approach for this area, through which a critical questioning of already existing theories is made possible and in which an important achievement is made both for their confirmation and for their invalidation. For example, Guttmann (2000a, p. 85) stated that:

> "if for a construction, which was built solely on data of one level and was initially exclusively related to it, in the other area a complete correspondence of the classes, relations and laws can be recognized, then it is obviously distinguished by it before all innumerable other possibilities, of which none will show this correspondence."

So, if another scientific discipline with other methods delivers the same result in the same question or topic, the probability of the content being valid increases compared to those who find knowledge based on only one method. For this reason, the findings found in the neurosciences are included not only because of their progress, but also because of the connection of propositions with other disciplines.

In the course of understanding the emergence of thinking, insights from the field of perception processing are used to derive both fundamental functions of neuronal circuits and to draw conclusions about the emergence of thoughts, connections, and associations. In this area, findings from animal experiments as well as from human experiments are included.

An important approach for the author is the work with model neurons (cf. Rösler 2011, p. 29f.), which are able to simulate the interconnection principles (Rösler 2011, p. 25) and thus can compare models such as Anderson's ACT-R with experiments from psychology or cognitive psychology, although from the author's point of view essential possibilities of nuancing through abstraction could be lost, if the related intermediate stages are not sufficiently taken into account.[3]

The author sees further important findings in this area in the development of (auto-) associative networks (Rösler 2011, pp. 30f., 36), which represent an

[3] The author tries to state that these models are suitable for simplification and examination. Nevertheless, it must be considered that temporarily extending fluent changes of nuances of excitations of neurons are considered only in the abstract. These fluent changes influence the repolarisation as well as the refractory phase. The latter, for example, causes a delay in continuous signaling. Missing this accuracy could lead to an information technology-influenced growing paradigm shift towards a digital, therefore discrete, process-related understanding.

adaption of neurons in a network connecting by associative linking. This makes them more stable against failures, which confirms one of the special characteristics in the brain, and at the same time they do not store details through these associations but are able to generalise patterns (Rösler 2011, p. 33). They can only do this by applying the principle expressed in Hebb's learning rule: "Neurons that fire together wire together" (Rösler 2011, p. 31).

This work is thus in the field of cognitive neuroscience, which is based on Miller and Gazzangia:

> "to bring together the more psychologically oriented neurosciences under a common term and to distinguish them from the more physiologically oriented neurosciences." (Rösler 2011, p. 6).

2.4 Psychotherapy Sciences

The different schools of psychotherapy sciences consider the treatment of "psychosocial or psychosomatic behavioural disorders and states of suffering" (Psychotherapiegesetz, §1 par. 1). Their goal is to achieve an improvement for the patient through the interventions. However, as these are not only aimed at curing illness, but also at promoting maturation and health, the findings from this work are useful here.

Psychotherapy sciences handle sick people. The development of mental illness can have many causes. It is often the combination of different factors, such as a genetic predisposition, early experiences in childhood, and triggering events (Roth and Strüber 2014, p. 297). Because psychotherapy works, which nobody doubts anymore today (Roth and Strüber 2014, p. 327), the changes produced by the interaction are an interesting aspect here, which is reflected particularly in a change of the oxytocin release as well as the stress reduction and the associated increase of neurogenesis (Roth and Strüber 2014, p. 359).

The extent to which the emergence of something new in this field is effective and important as an element for healing can be understood by analogy. "Our language is rich in images; they illustrate the world, make it comprehensible – and they do so by establishing a similarity relationship between the actual word and what it means." (Rieken 2010, p. 193). Change may require the basis of a hermeneutically reflected understanding of the world, which moves circularly via analogies to change.

2.5 Creativity Research

Creativity research concerns particularly extraordinary achievements, which are described as particularly creative by a peer group in the respective environment. Through this so-called *jury approach*, these limitations make case studies possible, with which both personalities with their character traits and the processes of these persons, who are often referred to as geniuses, can be examined (cf. Andreasen 2005, p. 29f.).

As the author focuses on the creation of something new, the process-oriented approaches, which on the one hand consist of subjective introspections and on the other hand of objective tests and experiments, are included in this work (Carson 2010, pp. 6, 57, 62; Andreasen 2005, p. 32).

By referring to the REST state of Andreasen, that is, the unconscious free flow of thoughts, and the connection with mental illnesses, derivations are made from neuroscientific experiments (Andreasen 2005, pp. 77, 80f.) to the models of thought.

A brief excursion concerns the relationship between creativity and intelligence, which is considered mutually limiting in connection with the threshold value (Smith 1971, abstract). The differences in performance between artists and writers and physicists in this area require more detailed consideration.

2.6 Constructive Realism

Because the work presented here is a theoretical examination of empirical findings, constructive realism is used to carry out a relativizing limitation.

The realisation that the observer is not independent of the observed events can be shown in many different ways. In quantum physics, it is already a fundamental problem that the observation of a photon influences its behaviour, resulting in seemingly paradoxical results. In some situations, the photon behaves like a particle, in others like a wave. Furthermore, as a result of the quantum-physical state of the entanglement of two particles, an observation of the first particle taking place simultaneously[4] occurs to the unobserved one and thus produces a different reality for the observer than if no observation had taken place (Zeilinger 2005, Chapter II. Neue Experimente, neue Unsicherheiten, neue Fragen).

The perception of the environment is another aspect that supports this theory. On the one hand, sensory signals of the retina, for example, are already reduced moments of reality: this is the result of the arrangement and interconnection of the sensory cells in the eye, which are connected in so-called on–off centres. The same combinatorial interconnection of cells is found in the visual cortex, so that the resulting interaction patterns, which are achieved by mutual excitation and inhibition, enable the recognition of contours, movement, and colour (cf. Rösler 2011, pp. 60, 70). In this case, however, this happens through a reduction of reality, ergo, through a predefined circuitry that constructs reality itself. This relativity also explains perceptual phenomena such as simultaneous contrast effects or successive contrasts (Rösler 2011, p. 58). Sensory filters are already changing reality. Therefore, there can be no objective reality, but only a subjectively created, that is, constructed reality.

[4] In the meaning of occurring at exactly the same time.

However, by recognising the dependencies of human beings in their knowledge of themselves and their actions, it is possible to take dependencies into account; the author sees radical constructivism as not sufficient, but even limiting, because it leads through exclusivity to total bondage.[5] Rather, in this respect he joins constructive realism, in which the relativity of knowledge in the existence of "Micro Worlds" is understood with the liabilities valid there, which make knowledge possible by alienating the constructed models and their connection between construct and prerequisite (Greiner 2012, p. 32f.; Greiner 2013, p. 4f.; Guttmann 2016, passim).

References

Books and Journals

Anderson, J. R. (1996a). *Kognitive Psychologie. Eine Einführung.* (2nd ed.). Heidelberg: Spectrum Academic Publishing House Ltd.

Anderson, J. R. (1996b). ACT. A simple theory of complex cognition. *American Psychologist, 51*(4), 355–365.

Anderson, J. R., Bothell, D., Byrne, M. D., Douglass, S., Lebiere, C., & Qin, Y. (2004). An integrated theory of the mind. *Psychological Review, 111*(4), 1036–1060.

Andreasen, N. C. (2005). *The creating brain. The neuroscience of genius.* New York/Washington, DC: Dana Press.

Carson, S. (2010). *Your creative brain.* San Francisco: Jossey-Bass, A Wiley Imprint.

Greiner, K. (2012). *Standardisierter Therapieschulendialog (TSD). Therapieschulen-interdisziplinäre Grundlagenforschung an der Sigmund Freud Privatuniversität.* Vienna, Sigmund-Freud-Privatuniversitätsverlag.

Guttmann, G. (2000a). Die Messbarkeit des Erlebens und der Irrtum mit dem Ich. In F. Stadler (Ed.), *Elemente moderner Wissenschaftstheorie.* (pp. 73–91). Vienna/New York: Springer.

Guttmann, G. (2016). Erfinden statt entdecken. In N. Holzenthal (Ed.), *Constructing reality/ Realität konstruierend* (pp. 25–32). Frankfurt am Main: Peter Lang Publishers.

Kahle, W. & Frotscher, M. (2005). *Nervensysteme und Sinnesorgane.* In: Taschenatlas Anatomie. 3rd band (9th revised edition). Stuttgart. Thieme Publishers.

Psychotherapiegesetz. BGBl. Nr. 361/1990 as amended on. Wien.

Rieken, B. (2010). Das Analogiedenken und seine Bedeutung für Medizin und Psychotherapie. Potsdam. AGEM (Arbeitsgemeinschaft Ethnomedizin e.V.). *Curare,* 33, 192–199.

Rösler, F. (2011). *Psychophysiologie der Kognition. Eine Einführung in die Kognitive Neurowissenschaft.* Heidelberg: Spectrum Academic Publishing House.

Roth, G., & Strüber, N. (2014). *Wie das Gehirn die Seele macht* (5th ed.). Stuttgart: Klett-Cotta Publisher.

Silbernagl, S. & Despopoulos, A. (2003). *Taschenatlas der Physiologie* (6th corrected edition). Stuttgart: Thieme Publishers.

[5]According to this, Wallner states that the radical constructivism insight, that everything is a construction, is unsubstantial, because it is too unspecific/universal (Wallner 2002, p. 183). Moreover, sciences referring to this statement would draw back to a simple data collecting something without any objective connectivity, because of the missing guidance of the statement. It therefore would lead sciences to an instrumentalised and technologically simplified state of art (cf. Wallner 2002, p. 189).

Smith, I. L. (1971). IQ, creativity, and achievement: interaction and threshold. *Multivariate Behavioral Research, 6*(1), 51–62.
Wallner, F. (2002). *Die Verwandlung der Wissenschaft.* Jandl, M.J. (Hg.). Hamburg: Dr. Kovač.
Zeilinger, A. (2005). *Einsteins Schleier – Die neue Welt der Quantenphysik.* Munich: Wilhelm Goldmann Publishers.

Other

Greiner, K. (2013). *Konstruktiv-realistische Psychotherapiewissenschaft.* Vienna/Paris: Teaching Script of the SFU Sigmund-Freud-Privatuniversität Wien/Paris.

Chapter 3
State of Research

Contents

In this chapter, the current state of the art of science in relation to the relevant disciplines is reviewed.

3.1 Anatomical Basics

This chapter refers to the functioning and the crosslinking of cells, as well as basic aspects of the brain-specific structure, and further on leads to the possible adaptation mechanisms in these areas.

3.1.1 Fundamentals of Cell Level

The nerve tissue consists of nerve cells and supporting cells, whereby the nerve cell, the ganglion cell or neuron, represents the actual functional unit and is supported by the glial cells (cf. Kahle and Frotscher 2005, p. 18).

Impulses can be transmitted between the neurons both electrically and by signalling substances, as occurs at the synapses, which form the contact point between two neurons. With a few exceptions, these synapses have a valve function from the transmitter–receptor effect, as they primarily transfer in one direction (cf. Silbernagl and Despopoulos 2003, p. 42).

K. Testor, *Cognitive Theoretical Foundations of Creativity*,
https://doi.org/10.1007/978-3-030-61736-3_3

Synapses are the locations where signal transmission can be excited or inhibited by the interaction of various incoming signals, whereby the sum of these processes can also lead to a shift in one direction or the other; they can be adapted (cf. Silbernagl and Despopoulos 2003, p. 42).

The axon is the efferent part of the neuron, which originates at the so-called axon hill and can have branches, called collaterals. These collaterals end at the presynaptic end boutons. Dendrites are the excitatory, afferent, parts of the neuron (cf. Silbernagl and Despopoulos 2003, p. 42).

The synaptic transmission can take place both electrically and chemically, whereby the electrical variant is the less common one. In chemical transmission, neurotransmitters triggered by an action potential are released into the synaptic cleft and dock postsynaptically to the specific receptors where they excite (usually acetylcholine or glutamate) or inhibit (usually γ-aminobutyric acid [GABA] or glycine, or acetylcholine at the M2 or M3 receptors at M-cholinergenic synapses) (cf. Silbernagl and Despopoulos 2003, pp. 50, 52).

In synaptic transmission, a presynaptic depolarisation, which involves Ca^{2+} ions, stimulates the release of signalling molecules into the synaptic cleft. The substances emitted by the fusion of the vesicles with the cell membrane bind to the receptors specific to them and thereby open the Na^{+}–AMPA receptor inputs (AMPA, alpha-amino-3-hydroxy-5-methyl-4-isoxazolepropionoic acid). In the administration of a longer impulse, an additional blockade of Mg^{+} also leads to a Ca^{2+} influx. A more rapid depolarisation leads to a longer cell response and can also lead to a change in gene expression on the postsynaptic side (cf. Kahle and Frotscher 2005, p. 30).

Synapses basically have two types of signalling receptors. First are the ligand-controlled ion channels, which are responsible for the rapid synaptic potential but only last for milliseconds. Their excitatory effect is usually based on glutamate as a transmitter and ensures a high level of Na^{+} or Ca^{2+} influx or K^{+} defluxion for the positivisation of the resting membrane potential. On the inhibitory side, GABA or glycine inhibits the inflow of Cl^{-}, which triggers the hyperpolarisation but above all the short-circuit current against incoming exciting impulses. Second, these are G-protein-coupled receptors whose responses last from seconds to minutes (cf. Kahle and Frotscher 2005, p. 30) and represent the target receptors of hormones or neuropeptides.

Postsynaptic inhibition is not the inhibition of synaptic excitation transmission, but the inhibition of the discharge of the postsynaptic neuron by inhibitory transmitter substances. Inhibition can take place in three ways, always requiring an interneuron (cf. Kahle and Frotscher 2005, p. 34):

- Recurrent Inhibition

In recurrent inhibition, an interneuron is excited from a projection neuron, which is connected back to the projection neuron and then inhibits it when an action potential is triggered and maintains the inhibition for further excitation transmission beyond the refractory phase resistance.

- Forward Excitation

A forward inhibition emanates from a neuron and excites an inhibitory interneuron, which is connected to a projection neuron; this is inhibited by this circuitry for excitations.

– Disinhibition

In disinhibition, the circuit pattern inhibits inhibition. An interneuron excited by a neuron inhibits a downstream interneuron, which cannot thereby inhibit the projection neuron.

Because a single input of excitation is not sufficient for exceeding the threshold potential, the synaptic pathways are essential. This effect describes the fact that at a frequency of more than 30 Hz in the input, the Ca^{2+} cannot yet reach the rest value, which is given at the output level, but is initiated again by the incoming signal and thus a summation of the residual value and predepolarisation triggered by the new signal occurs. This increase in the imbalance in the direction of positive polarisation was achieved by the initiation of the first stimulus (cf. Silbernagl and Despopoulos 2003, p. 50). In this case, one speaks of the spatial (see Fig. 3.1) or temporal (see Fig. 3.2) summation of the excitation, which makes it possible to exceed the threshold potential.

In this context, it is interesting to note that more than one bouton of an axon can be connected to the dendrite of the connected neuron (cf. Bartol et al. 2015),

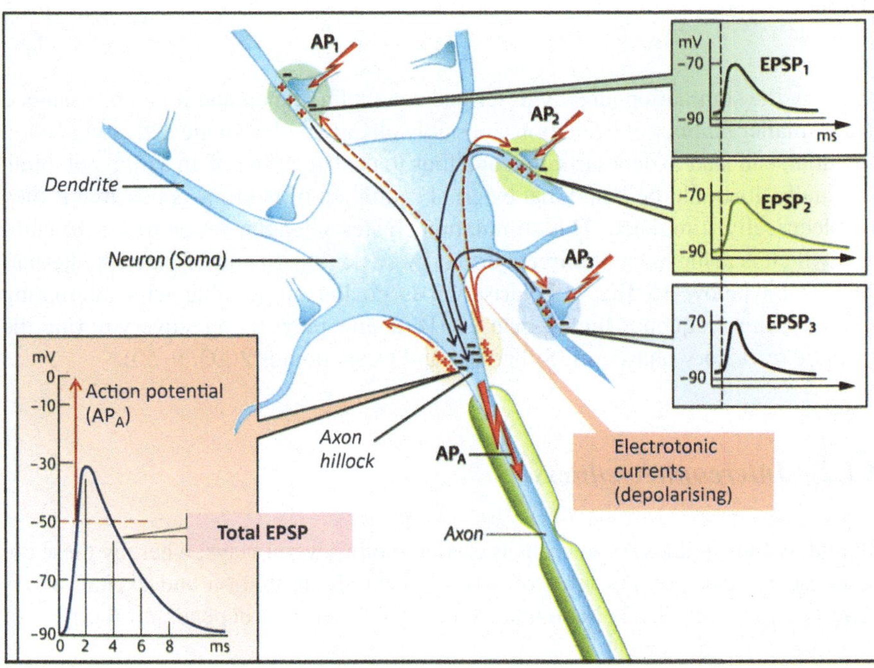

Fig. 3.1 Spatial summation of nervous excitation (cf. Silbernagl and Despopoulos 2003, p. 53, figure 2.6 B)

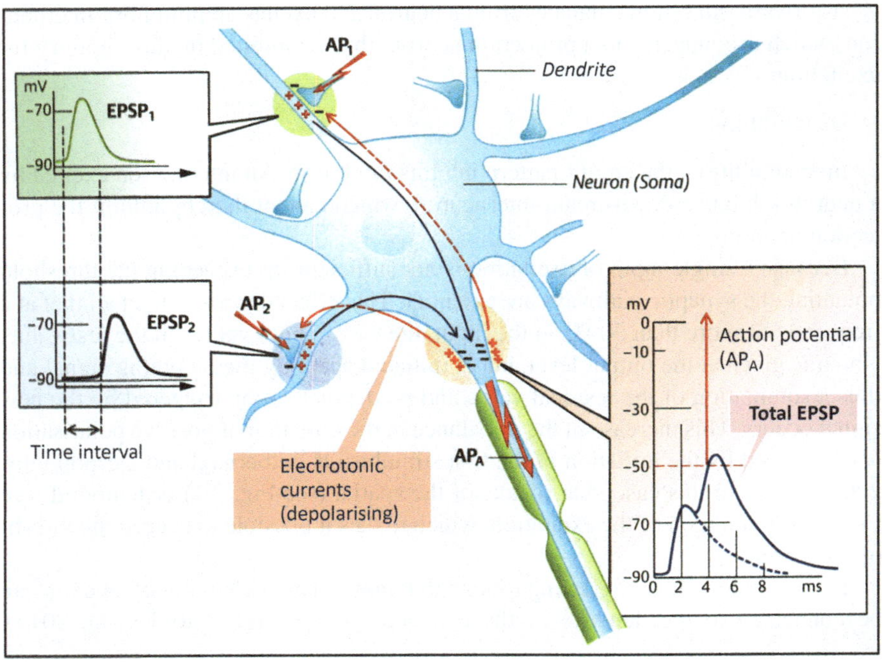

Fig. 3.2 Temporal summation of nervous excitation (cf. Silbernagl and Despopoulos 2003, p. 53, figure 2.6C)

whereby the summation already described can be fine tuned and it can be assumed that a smaller number of excitations, possibly already with a single activating action potential, can lead to depolarisation and thus to the triggering of an action potential.

Finally, it should be noted that external excitation of neurons is possible if they are electrically stimulated. This stimulation creates a depolarisation under the cathode, which is achieved with direct current by switching on and off or with alternating current below 15 Hz. Depolarisation is no longer possible with alternating current with a frequency higher than 15 Hz, as this prevents a recovery during the absolute refractory phase[1] (cf. Silbernagl and Despopoulos 2003, p. 50).

3.1.2 *Interconnectedness*

The interconnectedness of neurons is carried out in several ways, whereby these are based on the two possible uses of action potentials: inhibition and excitation. As mentioned in Sect. 3.1.1, Fundamentals of cell level, this depends on the type of

[1] Refractory phase is the phase after depolarisation and release of the action potential. During this phase neurons can hardly (relative refractory phase)/not (absolute refractory phase) release another action potential.

messenger substances that are released. Exemplarily, the following list gives a short overview of these network structure elements (cf. Rösler 2011, p. 25, Fig. 2.5; see Fig. 3.3):

- Convergent circuits combine many afferent neurons into one efferent neuron.
- Divergent circuits use the collaterals of the axon to only use one afferent neuron to excite several efferent ones.
- Presynaptic facilitations additionally integrate an excitatory interneuron initiated from another region to the afferent excitatory neuron, which thus sums the excitation potential both temporally and spatially.
- Presynaptic inhibitions function according to the same principle as presynaptic facilitations, but with the effect of inhibition.
- Inhibitory feedback circuits are connected such that the efferent neuron excited by an afferent neuron has a connection to an inhibitory interneuron that is retroactively effective on the efferent neuron and thereby complicates subsequent activation. Thus, the excited neuron inhibits itself for further activation by using an interneuron.

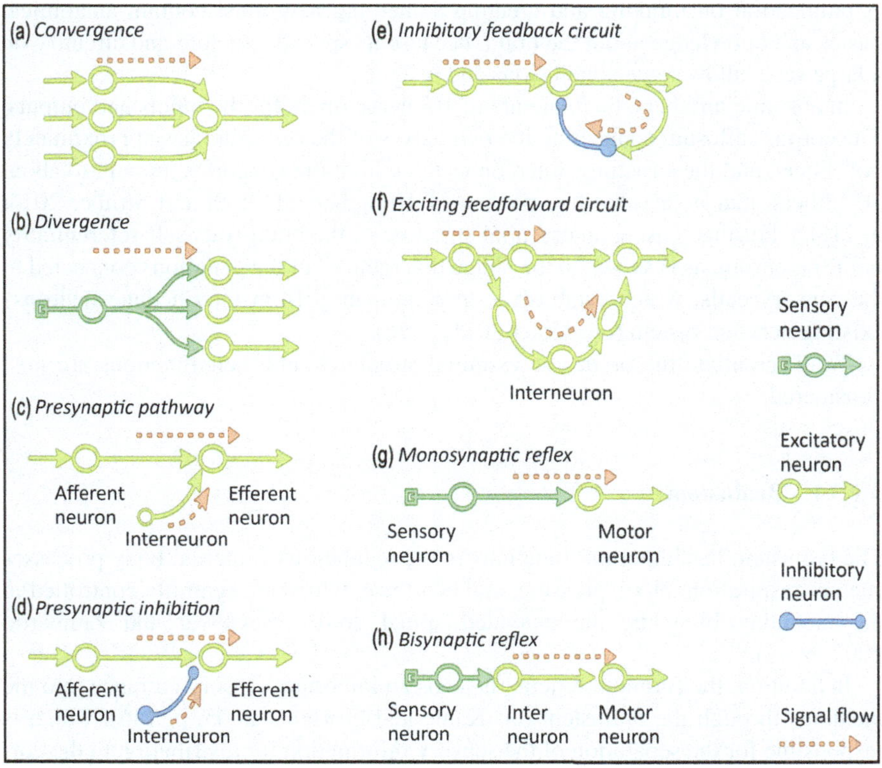

Fig. 3.3 Different kinds of interconnectedness of neurons (figure generated by author regarding Rösler 2011, p. 25)

- Exciting feedforward circuits are networked from the afferent neuron as a second path of orbit via one or more interneurons to the efferent neuron, resulting in a time-shifted amplification of the exciting signal.
- The monosynaptic reflex is the switching from a sensory neuron directly to a motor neuron, whereby, for example, the stretching of the leg can be triggered immediately in the event of a blow to the knee (patellar tendon reflex).
- The bisynaptic reflex in this case is the interposition of an interneuron into the monosynaptic reflex, whereby other influences can intervene in the course of the reflex by acting on the interneuron.

As is already seen from these circuit structures, the essential function of the neural network is that of a reinforcing and inhibiting network circuit, which can have a wide variety of properties from its different structure formation, which becomes even more complex when combined over several levels.

3.1.3 The Brain

A publication on thinking and creating something new must contain anatomical basics and knowledge about the brain, because its special structure and circuitry on a large scale allow some conclusions and insights.

In absolute numbers, there are about 10^8 nerve fibers for the inputs and outputs. The corpus callosum connecting the two halves of the cerebrum has approximately 10^{10} fibers, and the structures interconnected within the neocortex amount to about 10^{13} fibers, that is, a value five powers of 10 higher (cf. Roth and Strüber 2014, p. 218f.). For this reason, an essential structure of the brain is its self-referentiality in the cerebrum, as is shown in the "huge divergence" of the ganglions connected to the sensory cells, which establish a "predominantly from intermediate neurons" existing nervous system (cf. Rösler 2011, p. 26).

To differentiate the brain, the essential structures and their functions are now considered.

3.1.3.1 Brainstem

The brainstem has important functions in the regulation of internal body processes such as respiration, blood pressure, and heartbeat, which are centrally controlled in the medulla oblongata, the extended spinal cord (cf. Gerrig and Zimbardo 2007, p. 92).

In addition, the formatio reticularis leads an important sensory entrance into the thalamus through the brainstem (cf. Kahle and Frotscher 2005, p. 146), which is responsible for the separation of the sensory information for distribution to the corresponding areas in the cortex. The formatio reticularis is involved in attention control in both the waking and the sleeping state (cf. Gerrig and Zimbardo 2007, p. 92).

3.1.3.2 Cerebellum

The cerebellum is integrated into the base of the skull, lying on the brainstem, and is responsible for movement coordination, balance, and posture control (cf. Gerrig and Zimbardo 2007, p. 92; Kahle and Frotscher 2005, p. 152). Automated processes are already achieved by a control system emanating from the basal ganglia using the motor cortex and the cerebellum (cf. Roth and Strüber 2014, p. 212).

An important basis for the change in the interconnection is adult neurogenesis, that is, the formation of new nerve cells. This development also takes place in the cerebellum (cf. Carletti and Rossi 2008, abstract), which explains the motor and processual learning ability. In this sense, the cerebellum probably is involved in the adaptive capacity of the brain.

3.1.3.3 Limbic System

In the limbic system, which includes the amygdala, the hippocampus, and the hypothalamus, motivating behaviour and emotional and memory processes take place (cf. Gerrig and Zimbardo 2007, p. 92; Silbernagl and Despopoulos 2003, p. 330).

Connectivity studies show that medial and orbitofrontal limbic cerebral cortex areas are projected into the amygdala, but the dorsolateral prefrontal cortex hardly does so, which is why the emotion arising in this area dominates rationality and not vice versa (cf. Roth and Strüber 2014, p. 348). Because of this interconnection structure, the consciousness of these processes and their influence is hardly noticeable (cf. Roth and Strüber 2014, p. 228).

In this context, the amygdala is associated with emotional control and with the formation of emotional memory (cf. Silbernagl and Despopoulos 2003, p. 336; Gerrig and Zimbardo 2007, p. 93). But it is also often seen as a fear memory or fear organ, because the following applies: "The amygdala won't forget!" (cf. Roth and Strüber 2014, p. 366), which is particularly true when it comes to strong stresses and memories of trauma. It is important to note that the amygdala is involved in all events and can thus influence behaviour, although some signals are too weak to become conscious, but nevertheless trigger emotional or behaviour-influencing changes through the thalamus excited by subcortical processes induced by activations of the amygdala (cf. Roth and Strüber 2014, p. 201; Silbernagl and Despopoulos 2003, p. 330).

The amygdala, together with the central grey matter, which is also a substructure of the limbic system, has an influence on the distinction between important and unimportant stimuli. These structures are seen as jointly responsible in the assessment of the importance between the N100 and N400 waves of the event-related potential (cf. Roth and Strüber 2014, pp. 66f., 212).

The hippocampus is an essential structure for memory. It is associated with explicit memory and the acquisition of knowledge associated with it (cf. Gerrig and Zimbardo 2007, p. 92f.) and can lead to long-term anterograde amnesia in the event of damage, as has happened in the case of its removal for the treatment of epilepsy following a necessary intervention. The learning ability of this patient, known by the acronym H.M., was limited to the psychomotor abilities in which progress could still be made: this was not the case in all other learning areas (cf. Rösler 2011, pp. 177f.).

In this context, the hippocampus is understood as a memory for cortical addresses, which forms a compressed address code as a reference to memories in the cortex, where a (relatively) complete storage of excitation patterns, stimuli, context references, etc., is then available (cf. Rösler 2011, p. 195). For this consolidation of knowledge, the hippocampus also has an important function in sleep (cf. Rösler 2011, p. 208, in conjunction with pp. 205f., 211; Schandry 2011, p. 385), in which it connects the temporal sequence via spatial arrangement, which introduces the integration of the place cells and thus their contexts as an important aspect (cf. Roth and Strüber 2014, p. 82). The neurogenesis taking place in the hippocampus (cf. Roth and Strüber 2014, p. 253f.) is held responsible for the implementation of the new experiences and thoughts. This arrangement was confirmed by studies on rats by Kitamura et al. (2009) insofar as inhibited neurogenesis could show the prolonged presence of an unpleasant association with a stimulus in the hippocampus. The stimulus led to the disappearance of this association by a drug switching off the hippocampus, which was not the case in comparison groups (cf. Kitamura et al. 2009, abstract).

The hypothalamus consists of many nuclei, "which regulate the physiological processes of motivational behaviour" (Gerrig and Zimbardo 2007, p. 93). It controls many vegetative regulations and influences higher brain performance just as it is responsible for the sleep–wake rhythm (cf. Silbernagl and Despopoulos 2003, p. 280). It is supplied with much sensory and humoral information, that is, information concerning bodily fluids. It achieves control through both the neuronal circuits and the influence on endocrine processes of the pituitary gland (cf. Silbernagl and Despopoulos 2003, p. 330).

Another performance of the hypothalamus, in conjunction with the other parts of the limbic system, is the manipulation of the remaining brain areas via "the monoaminergic systems (dopaminergic, serotonergic, noradrenergic neurons), which move from the brainstem to almost all parts of the brain" (Silbernagl and Despopoulos 2003, p. 330). The neuromodulators, neuropeptides, or neurohormones released in these areas have a longer duration of action and change the receptors located in the membrane of the synaptic contacts or inside the cell in such a way that a local effect can last from seconds to hours (cf. Roth and Strüber 2014, p. 95).

3.1.3.4 Thalamus

The thalamus is an area of the brain at which many connections of the cerebral cortex innervate (cf. Roth and Strüber 2014, p. 218) and is associated with the control of the execution of production processes (cf. Anderson et al. 2004, p. 1038). It is functionally "a transit station through which almost all afferent information from the periphery to the cerebral cortex and, conversely, all efferent information from the cerebral cortex to the periphery is passed through and switched over." (Rösler 2011, p. 105).

Because the thalamus channels the incoming sensory signals and subsequently forwards them to the affected areas in the brain (cf. Gerrig and Zimbardo 2007, p. 92), the gate function present in the thalamus is of interest, which it achieves by means of internally inhibiting or exciting circuit structures. In the thalamic reticular nucleus, for example, retroactive projections are achieved by interneurons, which produce an activation pattern similar to a shift in attention; this enables relevant information to be reinforced and irrelevant information to be filtered (cf. Rösler 2011, p. 106).

In addition to this filter function, information in the thalamus is passed on, as is the case with visual stimuli: these are transmitted through the thalamus on the actual visual path into the visual cortex, but on a second path parallel to the amygdala (cf. Macedonia and Höhl 2012, p. 84f.), which triggers a corresponding fear reaction and can thus already lead to a reaction before conscious detection (cf. Roth and Strüber 2014, p. 214f.).

3.1.3.5 Cerebral Cortex

The cerebrum is the largest part of the brain, in terms of both mass and volume, and is divided into four larger areas on the right and left hemispheres: the frontal lobe, the parietal lobe, the temporal lobe, and the occipital lobe (cf. Gerrig and Zimbardo 2007, p. 93f.). The brain has histologically distinguishable areas, as published by Korbinian Brodmann in 1909 in *Comparative Localization Theory of the Cerebral Cortex*, categorised into 52 fields. This differentiation, which is still known today as Brodmann–Areale (BA), has a not absolute but nevertheless striking connection with the functionality and specialisation of regions in the brain (cf. Macedonia and Höhl 2012, p. 48f.; cf. Roth and Strüber 2014, p. 60f.).

As already indicated, differentiations are also made in the cerebral cortex according to functional aspects. Of particular importance here is the visual cortex, which is located in the occipital lobe and is responsible in the first levels for the integration of contrast/depth, colour, shape, and movement of the visual stimuli (cf. Rösler 2011, p. 70).

Auditive information is processed in the temporal lobe, speech especially in the Broca and Wernicke areas. Further specialised areas include the motor cortex, which is located exactly in front of the sulcus centralis in the frontal lobe. The somatosensory cortex then follows posteriorly, whereby these are similar in referencing to the respective body regions (cf. Gerrig and Zimbardo 2007, p. 94f.).

As already mentioned at the beginning of Sect. 3.1.3, The brain, the cerebrum is characterised by the interconnection structures primarily as a self-referential system. Therefore, "the major part deals with the interpretation and integration of information," which is likely to take place in the so-called association cortex, whose areas are distributed over the entire cortex (Gerrig and Zimbardo 2007, p. 95). This distinction is still important for further work in the creation of new things.

Because this function is distributed over different areas of the cortex, it can also be seen that although different specialisations of areas exist in the brain, they do not work alone, but interact with others and also partially overlap. The motor and somatosensory interaction should be cited here as an example. When activity patterns are generated in the motor cortex in the area of the fingers for gripping an object, they are associated in the somatosensory area at the moment of gripping by the information about the temperature, touch, and surface of the gripped object, which is perceived by the fingers: this leads to interactions in both areas. Therefore, it is not surprising that the areas of finger movement are adjacent to the area of sensory information coming from the fingers. Also, the interconnection structure makes it understandable that these areas are not isolated from each other, but are connected. So, the various specialised networks work together and interact with each other in their processing.

3.1.4 Modification of Neuronal Interaction

Because the neurons represent the actual functional unit of the nervous system and exert a considerable influence on the entire system in their interconnection, an interesting question is how and when the interaction between these units can change. These changes are recognisable both in density and structure, as, for example, an investigation of London taxi drivers and their ability to orient themselves showed an effect in neuronal change. The density in the area of the hippocampus showed strongly increased values for better orientational skills (cf. Maguire et al. 2006, p. 1099).

At the level of anatomy, the question arises as to how these changes come about and what effects they can have. The adaptive processes of the brain are summarised under the generic term *neuronal plasticity* and are the basis for understanding learning and memory. Plasticity can occur as structural plasticity, neurogenesis, and synaptic plasticity (cf. Seifert 2009, p. 8). At the level of anatomical changes, synaptic changes, which can have the effect of long-term potentiation (LTP) or long-term depression (LTD), are of closer interest here.

3.1.4.1 Long-term Potentiation

The changes in the nervous system, which can last for hours or even days, were found by Terje Lomo and Timothy Bliss in 1966, when Lomo could permanently change the properties of hippocampal neurons by electrical stimulation (cf. Rösler 2011, p. 160). For this purpose, a path of the hippocampus was excited by a tetanic stimulus[2] and the response was compared in a target area before and after the tetanic

[2] A tetanic stimulus is an electrically induced stimulus with a high repeating frequency.

stimulus. The stimulus response changed after the electrical stimulation and was increased in strength for a longer period of time, that is, potentiated. The change lies in the high-frequency excitation of the cell, which causes a permanent change in synaptic efficiency (cf. Rösler 2011, p. 161f.).

The change occurs when the NMDA threshold is exceeded. Synaptic transmission functions essentially through presynaptic depolarisation, which causes Ca^{2+} ions to flow in, causing specific vesicles to fuse with the cell membrane in the active zone so that they can emit the transmitter into the synaptic cleft. These then bind to the receptors specific to them and open the Na^{+} inputs of the AMPA receptors. If the impulses last longer, a high-frequency impulse setting causes glutamate to dock to the NMDA receptors, causing an otherwise blocking Mg^{+} release that makes Ca^{2+} inflow possible. In these cases, the G proteins lead to a longer response of the cell, which is also reflected in a change in gene expression on the postsynaptic side (cf. Kahle and Frotscher 2005, p. 30).

Thus, two factors are necessary for a change in synaptic efficiency through LTP. First, glutamate must have been released from the presynaptic area in sufficient quantity, and second, the depolarisation must already have shifted towards an excitatory postsynaptic potential (cf. Rösler 2011, p. 163).

The LTP is further divided into an early and a late phase, whereby the early phase increases the efficiency of the synapse for 2–3 h by making other signalling molecules more sensitive to glutamate, and in an as yet unexplained way increases the probability of successful vesicle delivery on the presynaptic side. This is all done without changing the gene transcription, the late phase, which is necessary to change the LTP for days to weeks. In doing so the signalling pathway is altered up to the cell nucleus, where it leads to changes in the synthesis "of mRNA and proteins as well as a change of the postsynaptic cytoskeleton in favour of new receptor accumulations and larger postsynaptic zones" (Seifert 2009, p. 15; cf. Kahle and Frotscher 2005, p. 30 in conjunction with Silbernagl and Despopoulos 2003, pp. 336, 42, 10, 12).

Long-term depression has the same effect with reduced activity and thus makes a corresponding stimulus response in the postsynaptic cell more difficult.

3.1.4.2 Neurogenesis

Adult neurogenesis is the formation of new neurons. For a long time, the brain was said to be unable to perform this task. However, as we know today, this ability is present, at least in some parts of the brain.

Adult neurogenesis is primarily associated with the integration of new information into the existing network, which is why it also has an important role in adults. It occurs mainly in the subventricular zone and in the hippocampus (cf. Roth and Strüber 2014, p. 253f.). Learning has a positive effect on the integration of newly formed nerve cells, which otherwise die off (cf. Shors 2009, pp. 49, 51).

In the course of this integration of the new into episodic memory as well as its integration into other areas, the striatum shows itself as an important entry structure into the basal ganglia. It is therefore not surprising that the striatum is also able to generate new nerve cells (cf. Roth and Strüber 2014, p. 254f.).

Another factor for growth is the *brain-derived neurotrophic factor* (BDNF), which is said to have an important role in anatomical modulation as well as in the stimulation of neurogenesis (cf. Lee et al. 2002, abstract), and serotonergic receptors (5-HT_{1A} and 5-HT_{2B}) have a stimulating role in their excitation (cf. Roth and Strüber 2014, p. 256).

The transport of new neurons is based, among other things, on the blood vessels to transport, for example, the precursor cells of oligodendrocytes from the area of origin, the ventral ventricular zone, along a "signaling pathway" that is oriented towards these vessels to the goal (cf. Dejana and Betsholtz 2016, p. 341 in reference to Tsai et al. 2016). Tsai and colleagues associated the binding to the blood cells with the Sdf-1 molecules[3] expressed by them, which also suppresses the further cell differentiation of these oligodendrocyte starting cells (cf. Tsai et al. 2016, p. 382). To what extent other substances are involved in this process is an interesting question, but it can probably be answered by the influence of the ion equilibrium in the respective cell regions. This was also suggested by a recent study, which found this to be the case in embryonic development in mice in relation to Ca^{2+} activity (cf. Rash et al. 2016, p. 1).

Another aspect of neurogenesis is the mentioned dying of unneeded new neurons. The probability of cell survival in the hippocampus increases with the difficulty of the tasks to be performed, as shown in an experiment with rats. The formation of new nerve cells is therefore necessary for the learning of complex processes and not for the learning of new, simpler ones (cf. Shors 2009, passim).

At the same time, hippocampal neurogenesis is held responsible for storage in explicit memory. On the one hand, this depends on beneficial factors, but can also be reduced by disturbing factors (cf. Kitamura et al. 2009, introduction). An important aspect for the integration of new information is the limitation of stress hormones, especially cortisol. Under stress, however, the regulation of the corticotropin-releasing factor emanating from the hippocampus is reduced, resulting in increased circulation of such substances (cf. Roth and Strüber 2014, p. 228). The regions that regulate stress, including the hippocampus, react to this stress with long-term consequences or reactions (cf. Stankiewicz et al. 2013, p. 76), which can have a negative effect on the ability to integrate. Because adult neurogenesis is jointly responsible for the integration of episodically new experiences into the existing life history and stress activation is associated with a high cortisol release, which reduces adult neurogenesis, it is assumed that this could be one of the causes of mental disorders. This reduction could cause a problem by decreasing the lack of linkability in connection with a simultaneously occurring overgeneralisation of similar events (cf. Roth and Strüber 2014, p. 272).

[3] Sdf1 (stromal cell-derived factor 1) is a body natural signalling molecule, which is among other things associated with directing cell movement.

One can also increase the quantity of neurons produced by neurogenesis. An increased release of oxytocin reduces the level of cortisol and stimulates neurogenesis in the hippocampus and striatum, resulting in a better ability to integrate new substances, which is positive for the psychotherapeutic process, for example (cf. Roth and Strüber 2014, p. 359). A further possibility to stimulate neurogenesis is exercise (cf. Kempermann et al. 2010, pp. 1, 6), whereby in this context the BDNF could again serve a function, which is increasingly produced with greater physical exertion (cf. Ferris et al. 2007, conclusion).

In addition to the aforementioned reduction by stress, alcohol abuse can also have a negative effect on neurogenesis and the hippocampus, which is important for the integration of memory content (cf. De Bellis et al. 2009, discussion). Overall, there is an individual range of neurogenesis, which is given in a certain spectrum of natural variance, depending on the genetic predisposition (cf. Kempermann et al. 2006, p. 780).

3.1.5 Recapitulation

In summary, it can be said that the brain is a network of nerve cells. At the synapses, these can be exciting or inhibiting. The type and frequency of use changes the synaptic connection in the short or long term. But changes also happen through the replication and integration of new ganglia, resulting in a changeable, that is, plastic, structure on basis of the anatomy, which has coined the term neuroplasticity. Furthermore, it was found that the neurons can be distinguished histologically. This distinction at cell level is partly reflected in functional associations. These functional associations are areas or networks specialised in certain tasks, which are not active alone, but nevertheless primarily in the respective specialisation. This interconnection structure leads to a self-referential brain, which thereby serves the integration and association of information.

3.2 Neuronal and Cognitive Processes

In the cognitive sciences, experiments have collected knowledge about the nature of processes in the brain, which has subsequently been linked to basic anatomical knowledge and has become verifiable by the newer methods of neuroscience. In this chapter, essential aspects for the understanding of the development of processes in the brain are considered.

To approach the processes of thinking from the sensory recording to its processing, the subsequent contextualisation and storage, association, and retrieval of information, the first step is to outline the process of perceiving stimuli on the basis of the visual system.

3.2.1 Visual Perception

In visual perception, the light rays of the visible wave spectrum entering through the pupil and impinging on the retina are converted into electrical stimuli by the photochemical reaction taking place in rods and cones. This activation is switched immediately by downstream ganglion cells in such a way that on–off centres are created. Through lateral inhibition, that is, an inhibition of the environment, they achieve the intensifying contrast of the incoming signals (cf. Rösler 2011, p. 60). As in Fig. 3.4 (cf. Zsynth 2006), both excitatory and inhibitory stimuli are transmitted to the bipolar cells on the second level by the different activation strengths of the sensors: these sum up the excitations and inhibitions coming from the neighbouring cells and thus lead to an increase in contrast. The summation shown in the last line has increased the difference with respect to the peripheral stimuli at the beginning.

The optic nerve runs from the retina to the optic chiasma, where it continues as the optic tract (the nerves in the optic chiasma are split into the visual field, so that those coming from the nasal retinal half cross and the temporal nerves continue to proceed on their side), passing the thalamus and from there via the optic radiation in the primary visual cortex V1 (cf. Silbernagl and Despopoulos 2003, p. 358). In this area, probably by the same system of lateral inhibition, a kind of recurrence of the on–off centres from the retina is created, whereby the first forms can be

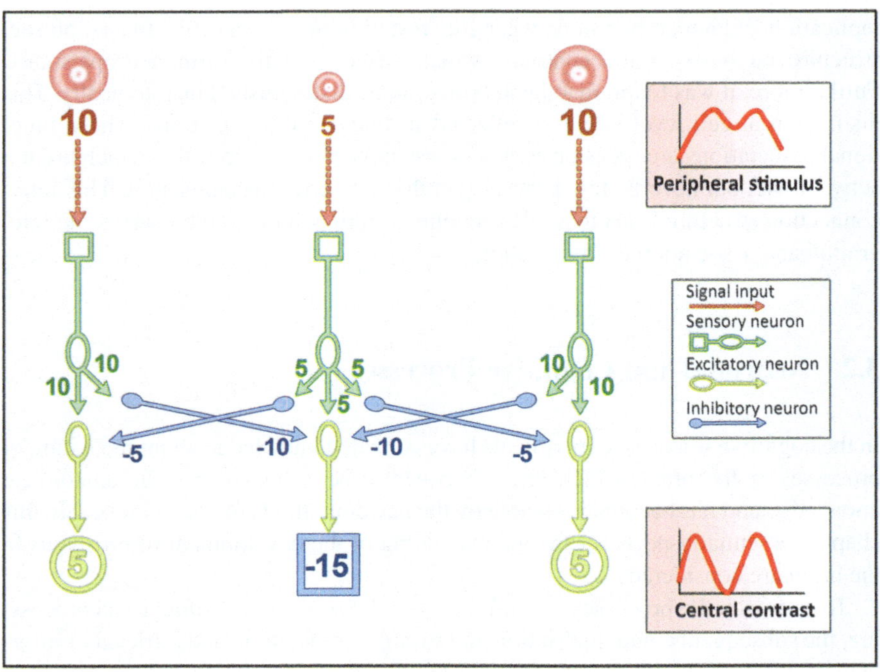

Fig. 3.4 Schematic picture of the effect of on-off centres (generated by author regarding Zsynth 2006)

generated here: short light stripes (cf. Macedonia and Höhl 2012, p. 95; Rösler 2011, p. 61). It is assumed that the signals arriving from the thalamus in V1 are converted by a combination of several receptive fields into a *simple cell*, the forms mentioned earlier. These rectangular receptive fields are both location-[4] and direction sensitive (cf. Rösler 2011, p. 59). Further connections of these "simple cells" then lead to higher constructs, the complex and highly complex cells, whereby it is assumed that the principle of interconnection continues (cf. Rösler 2011, p. 61). These cells, which specialise in alignment, perform an idealised filtering function ("tuning curve") because they can respond to the alignment in graduated activation patterns (cf. Rösler 2011, Fig. 3.8), and thus allow contrast with consideration of nuances (cf. Rösler 2011, Fig. 3.12).

This interconnection, based on microanatomical structure, is not yet fixed at birth and is only learned and developed through the experience of environmental stimuli. In 1973, von der Malsburg carried out a simulation using model neurons, which led to a combination of signals, initiated by standard stimulation and corresponding to the model of visual sensory lateral inhibition, to a pronounced representation of the stimuli after 100 repetitions. These beam receptors also made it possible to nuance the tuning curve and also reacted with their neighbours to similar stimuli (cf. von der Malsburg 1973, passim). This simulation, in which no special orientation was specified at the beginning and whose linkage weights necessary for the learning process were random, shows how filters can form through pure experience in a connective functional system (cf. Rösler 2011, p. 169).

The impact of visual deprivation was already studied in the 1970s. Experiments showed that in cats whose eyelids had been sutured after birth, there was no specialisation of the neurons in the visual cortex. As comparison group normally grown-up cats were used, in which from the fifth week an already adult cat corresponding adaptation had taken place (cf. Pettigrew 1974. passim). Thus, a genetically shaped structure was excluded.

Based on an experiment conducted by Blakemore and Cooper in 1970, in which cats grew up in an environment oriented towards purely vertical or horizontal environmental stimuli, developing only neurons with the appropriate preference orientation in the experienced environment, cats were subjected to early stimulus deprivation (up to 110 days after birth) with horizontal, vertical, or diagonal stimuli in a follow-up experiment. After a subsequent normal life under normal environmental conditions, the cats were examined after up to 3 years by uninvolved physiologists, veterinarians. It turned out that the stimulus structures experienced in the early weeks had caused a lasting influence on the directional sensitivity of the neurons in the cortex, so that these could be recognised afterwards by simply checking

[4]The location sensitivity of receptive fields of the retina is associated with the narrowly locally limited projection radius of exciting and inhibiting sensor cells. Therefore, the downstream ganglion cells are limited to exposure to light in a specified area in the field of vision. Through the interactions of these projections in the V1, these combinations form beam receptors also limited to specific areas in the visual field.

a stimulus response to different stimuli, although the cats moved and behaved relatively normally in the environment (cf. Muir et al. 1977, pp. 615, 628).

These studies were one of the reasons why the "developmental psychology" has been using "development window" or "sensitive phases," which assume a limitation of the acquisition of abilities as well as sensory performance in a specific time window, which, if not used, leads to a deficit that is difficult or impossible to make up. In this context, experiments on cataract patients who were exposed to lens opacity for a few weeks to months after birth until it could be removed by surgery may be mentioned as confirmation. In these patients, a limitation in contour perception can still be observed in adulthood (cf. Rösler 2011, p. 150).

This missing of critical phases, after which the learning of abilities would no longer be possible, was long suspected by doctors as irreversible. This presumption was disproved at some point by the project *Prakash*, initiated in 2004 by neuroscientist Pawan Sinah. In this project, Indian children who had been blind since birth because of a cataract were equipped with a synthetic lens instead of their own clouded lens (cf. Chatterjee 2015, pp. 372f.).

Manoj Kumar Yadav underwent surgery at the age of 18 after being blind from birth. Immediately after the operation it was possible for him to recognise light, but not to classify it. In the course of 1.5 years he was able to "see everything clear." This statement, which comes from himself, can be put into perspective insofar as he can lead a normal life and, for example, navigate through a crowd by bicycle, but he experiences limitations in his reading performance. With the help of glasses, he can read the headlines of newspapers, but he (still) has a hard time reading small print. Here, aspects of visual acuity as well as contrast sensitivity are not trained to the extent of normal development. Furthermore, he suffers from nystagmus of the eye, that is, uncontrolled movements of the eye (cf. Chatterjee 2015, pp. 372f., 375).

Thus, it is not possible for the treated persons to attain full sight, as already shown in the paragraph on developmental windows and sensitive phases, but nevertheless services are provided by the treated persons, which could no longer be explained with the original thesis of the missed period. The human brain obviously has a much higher plasticity than that of animals.

As an extreme case, birth blindness is interesting not only in the question of how the brain adapts to stimulation in later stages, but also what happens to unused areas during development. Studies of people born blind have shown that they begin to use the visual cortex for other specialisations by extending the areas activated for speech recognition to include areas of the visual cortex. Other studies have shown that tactile sensation also extends into these areas (cf. Rösler 2011, p. 154). A comparative study between early and late blind people showed that both groups showed increased activity in the visual cortex, at both the lower and higher levels. This activation was stronger in early blind people than in later blind people. In other areas of the brain there were no striking differences in activity (cf. Burton et al. 2002, abstract and discussion).

The distinction between lower and higher levels in a specialised field such as the visual cortex represents the increasing abstraction of sensory stimuli. Thus, at the level of the primary visual cortex V1 “highly specific feature filters that respond to retinal locations, line orientations and spatial frequencies” (Rösler 2011, p.74) are available. In the secondary visual cortex V2 there are “filters, which have very similar properties, but already ‘discover’ not only real edges, but also illusionary ones.” (loc. cit.). Here, as in the development of beam receptors, complex and highly complex shapes are made possible by progressive interconnections, which continues up to areas activated especially in the recognition of faces, the so-called *fusiform face area* (FFA) (cf. Rösler 2011, p. 74). Further areas are specialised in the recognition of objects and the dorsal as well as the lateral path process the meaning (“what”) and the information relevant for action (“how” and “where”) (cf. Rösler 2011, p. 76).

In one experiment, images of cars and faces as a whole, or in 2, 4, 16, or more parts, were shown in disassembled form, observing the activation in the visual cortex. On the one hand, it was found that the activation in the anterior and lateral areas of the area responsible for object recognition decreased with increasing division number; this was not the case with a halving of the image, either vertically nor horizontally. On the other hand, it was shown that activation in the upstream areas was not affected by this. The conclusion thus can be drawn that an increase in complexity in abstraction in the manner described here is represented by a hierarchy in the neuronal interconnection structure (cf. Lerner et al. 2001, abstract).

Another interesting aspect in this context is the question of when an object is assigned to one category and when to the other. As has just been stated, the object-related interpretation is carried out on the hierarchically higher levels. An experiment conducted by Labov in 1973 considered the question of whether containers for liquids are perceived as bowls or as cups. To this end, he created a wide variety of variants, which successively changed the form and blurred the uniqueness of the distinguishing features. Test persons preferred the designation “cup” up to a relative diameter of 1:1.9 (relation of height to diameter), whereas “bowl” did not occur up to this time at all, thus also other designations were used. At a higher relative diameter, both terms, “bowl” as well as “cup,” were used by 25% of the participants, which was called the blurring of these categories in this area (cf. Anderson 1996a. p. 155f). Combined with the previous findings, it can be stated that visual stimuli stimulated by similar to the same stimuli can lead to a competitive situation in interpretation. Lebov’s experiment in this case has an interesting resolution of this competitive situation: contextuality. In a second phase of the experiment he gave as additional information and task the instruction that the containers should be filled with mashed potatoes. By this simple change of the experimental arrangement, a shift of the object name to *bowl* already took place with a relative diameter of 1.2 in about 10% of the test persons and overtook with a relative diameter of 1.5 the number of designations as *cup*. The contextual reference has created a significant influence in this purely visual task (cf. Labov 1973).

This insight now leads to the question of how information is stored at all and how it can influence decision making through retrieval.

3.2.2 *Memory and Its Contents*

Experimental research into memory began at the end of the 19th century with Ebbinghaus, who had investigated the retention performance of senseless syllables after learning processes. The repeated recall gave him a forgetting curve that showed the percentage of forgetting over time in the absence of learning repetitions. In subsequent experiments, he showed that not only did the forgetting rate decrease with increasing repetition, but the repetition duration of learning also became shorter (cf. Anderson 1996a, p. 167f.).

Memory always involves encoding, retaining, and retrieving content, which is why the experimental distinctions between how information gets into memory and how it is retrieved are artificial (cf. Anderson 1996a, p. 168).

In the context of classical conditioning, Pavlov has shown that an originally neutral stimulus (NS), when combined with an unconditioned stimulus (US), becomes a conditioned stimulus (CS) after some time, which triggers the response to the unconditioned stimulus when the originally neutral stimulus occurs. So, in dogs the US "dog food" increases salivation. If now food and a NS "ringing of bells" occur together more often, the NS becomes a CS and the reaction of the salivary flow is already stimulated by the ringing of the bell, even if no food is provided (cf. Pavlov 2010 [1927]).

3.2.2.1 Neural Basics of Learning

In the context of anatomical basics, the potential for change of the synaptic connection was mentioned as essential for learning. The discovery of long-term potentiation is the confirmation of this idea: Hebb's learning rule, which means that the connection strength of two synapses changes with their experience (Hebb 2009 [1949] p.62):

> "When an axon of cell A is near enough to excite cell B and repeatedly or persistently takes part in firing it, some growth process or metabolic change takes place in one or both cells such that A's efficiency, as one of the cells firing B, is increased."

He thus anticipated the sustainable change in synaptic efficiency discovered in the 1970s in a theoretical sense.

High-frequency stimulation of nerve cells thus leads to an increased sensitivity of these cells. Barnes did so in 1979 by carrying out the procedure over a longer period of time and recorded the corresponding change. He discovered that the development of sensitivity, which he had measured by the growth rates of the excitatory postsynaptic potential, is a positive logarithmic function that seems to approach a physiological limit of information transfer (cf. Anderson 1996a, p. 186).

This finding is consistent to the extent that the adaptation of synaptic plasticity determined by Bartol et al. 2015 is possible through the multiple formation of boutons, but at the same time it is determined that the correlation of spine head volume and its neck diameter with both pre- and postsynaptic performance could lead to an anatomical limitation in the flow capacity of transmitter substances (cf. Bartol et al. 2015, Discussion of results). This would also explain the behavioural data obtained on the basis of anatomical limitations.

Hebb's learning rule is used in the simulation of learning networks to obtain a quasi-realistic instrument for model verification through the weighting and proximity relationships of the model neurons. The model neurons are excited or inhibited by others and thus calculate a sum of the states acting on them. In the course of the activations the linkage weights, that is, the aspects associated with the stimulus and thus learned, are changed. When two stimuli coincide, the linkage weight increases so that triggering can take place more easily with the next joint activation (cf. Görz et al. 2003, p. 84f.). Through this interconnection, a network learns independently which stimuli are connected to each other and which are not (cf. Rösler 2011, pp. 29–36). The impulse transmission from the model neurons changing in such a way proceeds according to the same principle as it is the case with neurons: according to the *all-or-nothing* principle. If the triggering threshold is activated, an action potential is generated (cf. Silbernagl and Despopoulos 2003, p. 46; Rösler 2011, p. 29f.).

Associative networks can be created by interconnecting several model neurons: these increase the signal-to-noise ratio through the learning experience (cf. Rösler 2011, p. 35) in such a way that this enables the filtering of triggering stimuli, as is the case with Pavlov's conditioning mentioned earlier. Here the connection between the signal bell and the feed is filtered out until the signal of the bell alone can trigger the saliva flow.

Associative networks "complete incomplete initial patterns, generalize, implicitly acquire the prototype of a stimulus set and are immune to injuries" (Rösler 2011, p. 37). These are connected in a network matrix consisting of two cell populations, so that one cell of the initial population can act on all cells of the target population. The link weights between the input and target cells can be different, which means that different output signals are generated from an input signal. A larger number of cells allows more variants to be mapped in the matrix and, as a result, a clear shift in the "association" (cf. Görz et al. 2003, p. 90f.). In auto-associative networks, the signals generated are presumably kept active for a longer period of time through the connection with themselves, that is, the feedback of the output signals, which is why this kind of system is thought as the theoretical foundation for the functioning of the working memory (cf. Rösler 2011, p. 37).

Another system of the learning model is the competitive network. This network is analogous to the associative network, but in contrast to this, it has switched declining inhibitory interneurons, so that it is able make use of "the statistical conditions (frequencies and correlations) of environmental characteristics and, on this basis, establish associations and classify them into categories" (Rösler 2011, p. 39).

3.2.2.2 Memory Systems

From personal experience we know different memorising and learning achievements, from short-term items such as names, numbers, or other information memorising situations to childhood memories or study and school achievements. The different categorisation of these time spans as different performance elements of memory, as in short-term and long-term memory, is understandable even if controversial.

The concept of short-term memory, which has been used in cognitive psychology for some time, following neurophysiological knowledge is no longer valid, but still contains important ideas (cf. Anderson 1996a, p. 169). One factor exemplarily can be mentioned for considering the recognition of limited validity: sensory content can reach long-term memory directly without requiring short-term memory (cf. Anderson 1996a, pp. 170, 82). In this context it can be shown that the sensory memory, also called ultrashort time memory, stores visual stimuli, for example, a complete visual arrangement, until 1 s after its presentation with a degressive tendency, that is, increasing loss. After 1 s, only that information can be remembered to which the attention was directed (cf. Rösler 2011, p. 175). The postexposure effect studied by Sperling in 1967 is also important in this context, because it showed that a bright field after a presentation of an image limited the aftereffect time to 1 s, whereas a dark field made it possible to retain the visual information for up to 5 s (cf. Anderson 1996a, p. 84). As a point of criticism, the necessary dwell time of information in an abstract construct of short-term memory is seen, which must be preceded by a storage: this does not correspond to the reality of perception. However, it must also be questioned here to what extent this differentiation as a concept thus only experiences a dimensional restriction and not the rendering untenable of the same, because thus indirectly the concepts of attention and awareness are regarded as necessary and not the question of whether there is an area of very short storage.

The contradictions that can be deduced from these findings, which speak against short-term memory, are contrasted by experimental indicators for different memory systems which show that information obtained in the short term is retained at a different depth than is long-learned information. There is, thus, a difference in the amount of information retained short term as compared with long term (cf. Anderson 1996a, p. 172). Baddeley tried to bridge this gap in verbal material with the construct of the articulatory loop. He also coined the concept of working memory, which for him had a dimension that could be filled with short-lived information, which, if no renewal occurred, faded from this memory after about 1.5 s. He deduced this from the relationship between word length and reading speed, which, with the same number of words, made a higher memorisation performance possible if the words had fewer syllables (cf. Baddeley 2003, p. 831f.; Anderson 1996a, p. 172f.).

Baddeley assumes a central executive as control, which controls these various auxiliary systems and can retrieve and feed information from them (cf. Baddeley 2003, pp. 830, 835). As an essential difference to the theory of short-term memory, information does not require a dwell time in the articulatory loop, which is only an auxiliary system for the availability of information and not a preliminary stage of storage (cf. Anderson 1996a, p. 174). A confirmation of this idea was given, for example, in an experiment by creating a "lexical bias," a sound or syllable exchange, which generated a burden on the central executive branch through an attention-intensive second task and thus disrupted the internal control process (cf. Hamm et al. 2004, abstract).

The central executive branch provides both prioritisations between search and evaluation of different areas from which memories are retrieved or which are to be processed from the sensory material (cf. Hills et al. 2010, abstract), as well as being necessary for the integration of activities from other areas, which would explain the further spatial dislocation (cf. Wiesner 2004, summary).

According to Baddeley, the central executive is thus a construct consisting of several elements, a phonological loop, a visual-spatial notepad, and an episodic buffer, which has also been shown in various studies.

Nevertheless, neurobiology distinguishes the construct of working memory from structural memory:

> "Structural memory content is formed by a permanent change in synaptic connectivity. The contents of working memory are defined by the number of activated filters, they are ultimately the result of reactivations and processes of attention control" (Rösler 2011, p. 211).

Thus, a second memory system comes to the fore next to the working memory, the structural system, also known as long-term memory, because its changes are structural adjustments.

Studies on learning concerned with the short-term retention of content focus on working memory. All others consider long-term memory, whereby the distinction between storage, retention, and retrieval is difficult to distinguish experimentally, because these processes are more or less part of the experimental setup in all cases (cf. Anderson 1996a, p. 193). In this context, it must also be taken into account that the distinction between working memory and long-term memory is difficult here as well. Memorisation tasks usually require a slightly longer period of time but can be influenced by the early phase of LTP, which is effective for 2–3 h in a change in synaptic efficiency (cf. Seifert 2009, p. 14). As this does not yet represent a sustainable structural change, the subdivision made earlier can still be made, but this line is just as poorly defined as the others.

Long-term memory comprises explicit (declarative) and implicit (procedural) knowledge. The former is the one that one can tell about and of which one is aware: the latter is knowing how to do something (cf. Anderson 1996a, p. S230).

The processing of information is performed in units of knowledge representation. These are general elements seen in the respective topic reference. For example, a chunk in the area of geometric figures can combine a sum of several more primitive units such as different bars into a triangle. Because chunks represent a generalisation of a principle, they can also assemble more complex figures from the triangles, such as a parallelogram or a diamond (cf. Anderson 1996a, p. 120). Through this combination of several chunks into a new unity of a representation of knowledge, a new one emerges, a new chunk. This action already results in a model of knowledge that is hierarchically related to one another, which can be changed depending on the relationship to other knowledge. In this sense, 'chunk' is also a simplistic abstraction, which enables a model design to be decoupled from the respective contents and facilitated but can still be tested again by recoupling.

For procedural knowledge, there is also a smallest unit of knowledge representation; these are production rules that enable problem solving as operators (cf. Anderson 1996a, p. 244). Production rules are created by the modification of chunks (cf. Anderson 1996b, p. 359), and so several production rules can be brought together by the production composition in such a way that the effect of both is produced in a new work, whereby the limitation to pure experience dissolves (cf. Anderson et al. 2004, p. 1045).

Long-term memory organises 'chunks' in both hierarchical and serial order (Anderson 1996a, S119–S125f.) and connects them in a hierarchical encoding of serially ordered information. This hierarchical storage occurs when large amounts of information are related to each other. An example of this distinction is counting with the aid of the hands. In Germany, children count one number per finger: that is, with one hand, 1 to 5, and with the second hand, 6 to 10. Chinese children count up to 10 with one hand. Thus, in the hierarchical order one hand is represented as the main constituent in Germany up to 5, in China up to 10. This distinction can be shown in a numerical comparison task. In this task, subjects receive a comparison task in which they are asked to decide which of two numbers shown is higher. Latency time measurement shows that in the comparison task between 4 and 6, a significant increase is recorded among Germans, which only occurs with 10 and 12 for the Chinese. The values 8–10 and 9–11, which are strongly below the rising curve, were triggered by the jump from single-digit to double-digit numbers. As can thus be seen, the curve is not only represented by the hierarchical order, but is also culturally determined, that is, learned (cf. Spitzer 2012, p. 174f.; see Fig. 3.5, p. 52).

A distinction of declarative memory concerns episodic and semantic memory. Semantic memory encompasses general (factual) knowledge and also includes language and general knowledge about the world. The episodic memory, on the other hand, contains experiences and moments that are connected with specific places at specific times (cf. Eysenck 2009, p. 200). The autobiographical memory is seen as a separate area, which is however connected with these (cf. Eysenck 2009, p. 206f.; Rösler 2011, p. 172).

The connection shown here in the declarative memory, which presents itself according to hierarchical as well as serial order, leads to another phenomenon, the so-called memory traces. These traces are to be understood as a construct radiating from an activated memory unit that enables or prefers an activation of the elements associated with this unit (cf. Rösler 2011, p. 176, Fig. 7.1; Anderson 1996a, p. 178).

Memory traces represent the course of fields and activations in the neuronal network. This network comprises two parameters: activation and strength. The activation is a sometimes rapidly changing scale; strength, however, changes only slowly. The strength is one of the decisive factors for how high the activation can be and thus also how accessible is the trace. The strength of a trace can be increased through practice (cf. Anderson 1996a, p. 183).

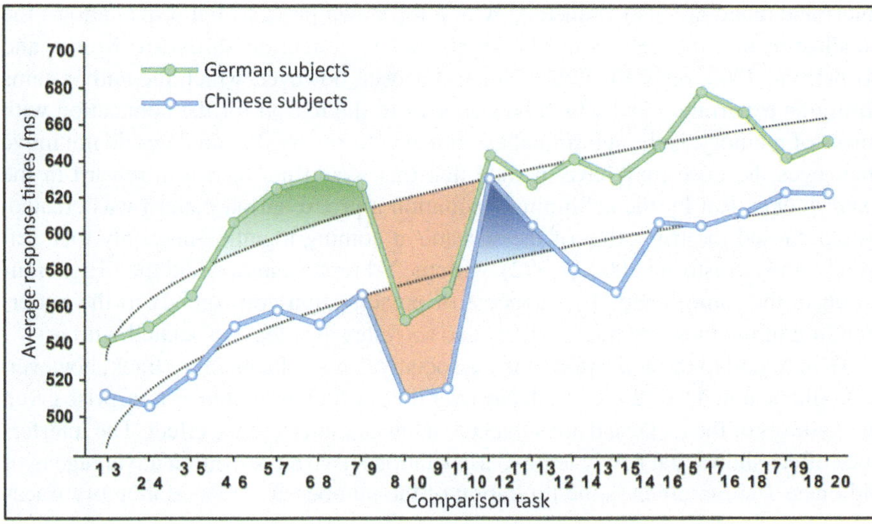

Fig. 3.5 Latency time differences of a number comparing task between German and Chinese subjects (generated by author regarding Spitzer 2012, p. 175 figure 7.9)

Traces of memory seem to decay over time, as is reflected in decay theory, which is a widespread theory of forgetting. In addition, this is also evident in studies over longer periods of time, which attribute a change after a certain duration to the circumstance of changed synapse circuits or densities (cf. Anderson 1996a, p. 199f.).

The memory traces are thus a preparation for an activation of related neurons, whose connection is associative in nature, as can be done by hierarchical as well as by serial order. Because these already radiate into other areas, a finding from an invasive experiment with macaques is interesting. In this experiment, macaques were taught to view a fixation point on a screen without looking away from it. However, after a surgical intervention, researchers were able to stimulate a specific area in the *frontal eye field* (FEF), which is also responsible for eye movement, so that short saccades as well as their direction could be triggered. In the experiment, however, a subliminal activation in a specific area was stimulated which in itself did not lead to movement (cf. Moore and Armstrong 2003, p. 371). In addition, the researchers localised a receptor field in V4, which overlapped with the area where the saccade would occur during an induced stimulation. For the experiment, activity was measured in the receptor field, which became active upon presentation of a preferred stimulus (*p*) but remained at baseline in a nonpreferred stimulus (*np*). The subliminal stimulation of the FEF area caused an increase in activation of the V4 receptor field; this effect was all the greater when an *np* was added to *p*. The interpretation is that subliminal saccade excitation projections contrasts with the *np* stimulus, reinforced by the fact that when *p* and *np* coincide with a subliminal excitation of the FEF area, which would stimulate a movement to *np*, the field in V4 for *p* is inhibited. Not only can it be deduced from this that fields V4 and FEF are interrelated and are also connected with a top-down process, but it also allows the conclusion that the FEF could be significant for attention shifts (cf. Moore and Armstrong 2003, pp. 370, 372f.). The last aspect, however, which the author gains from this experiment and which he also sees as the reason for the connection with traces of memory, is the subliminal excitement. Normally the monkeys do not move their eyes, because they have learned that this would not lead to a reward in the experiments. But by the subliminal excitation a preexcitation pattern was created, which caused the triggering of the saccade at coming together, and only then (cf. Moore and Armstrong 2003, p. 372), with the V4 representation and the FEF stimulation in the same range. This concept of preactivation corresponds to the understanding of the memory track, which also activates the areas associated with it.

With regard to the activation of the associative areas of a memory track, however, it should be noted that these can decay over time, which is no longer a surprise given the findings of the LTD, and are subject to a fan and interference effect. The interference effect shows that newly learned associations have an activation advantage over old ones. The fan effect is the limitation of the number of activated memory traces emanating from a 'chunk' (cf. Anderson 1996a, p. 201).

In the course of the memory traces and the preactivation, another phenomenon should be described briefly. Priming prepares the activation of areas associated by the triggering stimulus, prime. For this purpose, the activation should spread along the learned traces of memory and prestimulate the associated or obvious concepts. Priming takes time to prepare, reaching a maximum preparation after about 200 milliseconds (ms). After this point, the stimulus responses are not made more quickly (cf. Anderson 1996a, p. 182f.).

Speaking of associations, it is also necessary to take a closer look at conceptual knowledge. Schemes in this sense are "an ordered set of facts" (cf. Anderson 1996a, p. 213).

Schemes have a distorting effect when not experienced or unknown. They change memories, for example, the further away a story is from its own culture, which is reinforced by time that elapses over a longer period. Thus, Bartlett used a legend of an Indian tribe in 1937 to create an experiment. He presented his test persons with this tale, which was culturally and contextually extremely far from the probands' experiences, as was shown in the performance of the reproduction. With increasing time this story was changed by the addition and omission of details and information in such a way that it seemed to correspond more and more to the stereotypes existing in the person's own culture (cf. Anderson 1996a, p. 214f.).

Schemes not only represent a hierarchical level relationship, but also the relationship of properties to each other through a partial-overall hierarchy. They also allow a better comparison of terms, taking into account particular aspects. In this way, schemata can also be used to show what is common to things (Anderson 1996a, p. 150f.).

The extension of schemata to actions and events was carried out with the consideration of concepts for processes. In 1977, Schank and Abelson introduced the term scripts (cf. Anderson 1996a, p. 157).

Because schemes as well as scripts can distort not only perception but also memory, just as they seem to provide a framework for behaviour in certain situations, it is apparent to consider the influence of the environment on thinking, the context.

3.2.2.3 The Influence of Contextuality

Godden and Baddeley conducted an experiment with divers in 1975. For this purpose, a group of experienced divers from the university's own diving sports group was recruited as volunteers for a memorisation task. For these, a cassette was played to them from land via an underwater telephone, which first instructed a special breathing sequence to prevent underwater interference with the understanding of the recordings. Then the word lists were played twice on the cassette. The divers sat about 6 m under water on the ground and listened. To test the environmental influences, the same test procedure was carried out on land, whereby the divers also sat in a diving suit, but without wearing breathing masks and snorkels. Thus, the environment variants wet (w) and dry (d) were available for memorising. In this experiment, the query was made either in the same or in a different environment. Results of the combinations wet/wet, wet/dry, dry/dry, and dry/wet could be evaluated, showing the best performances were achieved under d/d, closely followed by w/w. The two environment change variants achieved only between 62% and 75% of the other performances, which showed a highly significant connection between learning environment and retrieval environment. This finding is a confirmation of the context-dependent memory hypothesis (cf. Godden and Baddeley 1975, passim).

In a metastudy on contextual memory, it was shown that the environment has a moderate but reliable influence on memorisation performance. This special example of Godden and Baddeley therefore is put into perspective. The metastudy also showed that the overshadowing thesis[5] that the environment cannot be stored, cannot be confirmed, because the focus is on other things: that is, that the other things would overshadow the environment, and therefore this can have no influence on the storage. In this context, it was found that while focusing during learning reduces the perception of the environment, this effect does not depend on time. This weakens the thesis to the extent that there should be a cumulative effect and thus the context effect should become stronger over time. Furthermore, the restoration of the mental state as well as conceptual processing were identified as relevant influencing factors (cf. Smith and Vela 2001, pp. 213, 215).

The context can also be created within another area, as shown by experiments that tie the memorisation of words with another word through contextualisation. Normally a recognition task shows better results than a reproduction task. The contextualisation of an experiment by Watkins and Tulving showed that the context can reverse the performance. For example, the test subjects achieved better results in

[5] The overshadowing thesis states that suppressing the environmental influence during learning prevents the storing of this additional information in the memory (cf. Smith and Vela 2001, p. 207).

reproduction, because the word to be reproduced was associated in the learning context. In the recognition task, however, the word, caused by an association task, was placed in a different context, which was reflected in a poorer value (cf. Anderson 1996a, p. 224).

From the synopsis it can be seen that context affects both the environmental stimuli and the mental state. The mental state also includes the emotional component. Posttraumatic stress syndrome (PTSS) or posttraumatic stress disorder (PTSD) are therefore likely to be a special form of context-related learning. These are caused by a particularly stressful situation or event and are reflected in increased agitation/alert readiness, avoidance of stimuli associated with the event, and the reliving of the events in nightmares or even flashbacks (cf. Eysenck 2009, p. 517). Although such a disorder is dependent on a bundle of factors, the influences of changes in the amygdala from a shifted abnormal hormone balance are often discussed on this topic (cf. Morey et al. 2012, conclusions), which could cause a predisposition, or, as discussed in other areas, a genetic indication in receptor omens can be recognised (cf. Xie et al. 2009, abstract; Kimbrel et al. 2015, discussion). Nevertheless, it is a disorder dependent on a triggering event.

In the treatment of such patients, attempts are made to forget what has been experienced, but this is not possible, as was explained during the decay of the memory traces. Only a nonactivation is possible, which causes an incipient decay of the memory track, but no "deletion." One approach is shown by an experiment by Manning, in which he combined a memorising and a forgetting task with scenic pictures with colleagues and recorded the areas activated for capture in a functional magnetic resonance image (fMRI). It turned out that those areas that used a reference to the scenery images in the presentation were less active in the forgetting task than in the memorising task. This finding shows that the thesis of reducing the recording of context leads to other contents being recorded less (cf. Manning 2016, discussion). This difference can also trigger the reduction of attention produced by conscious suppression, which acts as a mediator in the excitation of areas. In Sect. 3.2.2.4, Attention, this is explained in more detail. Nevertheless, even in this type of learning there is the problem that the experiences that have faded into the background can still be reactivated by a combination of other stimuli.

Derived from the shifted contextuality, a current experiment, used for the treatment of PTSD patients in the USA, makes sense. External stimulation of the trigeminal nerve[6] during sleep resulted in a sustained positive change in patient quality of life, stress disorder, and depressive symptoms within 8 weeks (cf. Cook et al. 2016, abstract). The connection is likely to be affected here, among other things, by the stimulation of the trigeminal nerve that distracts from the original contextuality,

[6] Nervus trigeminal is a somatosensoric facial nerve, which projects into the ventrobasal complex (p. 322). Its open nerve endings are triggered by aggressive olfactory stimuli and carry the information further to the bulbus olfactorius (p. 340). Additionally, this nerve is involved in the corneal reflex, which as a protective reflex coinitialises blinking (p. 359); all cf. Silbernagl and Despopoulos (2003).

although further experiments are needed to rule out any side effects and to deepen this insight.

By the recognition of the influence of the context on the processing, it was shown that the processing can be influenced by the conscious turning away, which brings attention as an accomplishment of the brain into focus.

3.2.2.4 Attention

Attention is a prerequisite for the sustainable storage of information. As has already been shown with sensory memory, its contents are retained for a short time, but after about a second it is only possible to remember that to which the attention was directed (cf. Anderson 1996a, p. 83f.). The processing and recognition of complex patterns and subtle changes requires full attention, as in these actions the working memory has to activate and process much information from the long-term memory (cf. Roth and Strüber 2014, p. 214). In addition to attention, emotional excitement has a role in the storage of information (cf. Roth and Strüber 2014, p. 227). However, this rather reflects the contextual framework mentioned in the previous chapter. What exactly does attention do?

In a video recording of two groups playing ball against each other, distinguished by black and white T-shirts, test persons should count the number of ball contacts of the white team and correctly name them at the end of the video. The video can be found, for example, on YouTube by searching for "selective attention test." In this 0:38-min-long film of the experiment conducted by Daniel Simons, tested persons mostly performed very poorly, but not in the counting task. In this video a person dressed as a black gorilla appears in the middle of the film, walks between the teams, even drumming his chest before it leaves the recording. However, this gorilla is not perceived by most of the test persons (cf. Simons and Chabris 1999, abstract).

Attention as a focus seems to influence what is consciously perceived from receptions. To better understand the mode of action, an experiment was conducted to investigate how neuronal activity changed to a visual stimulus through attention. In this experiment, a screen was used to suggest a figure appearing in one of four areas. Because this indication was that a stimulus should be evaluated for correctness, attention was paid to the appropriate area. During the experiment an event-related potential was recorded that documented the stimulus response of the neuronal activities: frontal, central, parietal, occipital, lateral occipital, and posterior temporal. It was found that neutral stimuli that appeared without a clue stimulus had a medium activation. On the other hand, a suggestion that drew attention to a specific area led to two different swings. In the case of a valid reference stimulus, the event-related potentials became stronger with the emergent figure, whereas it became weaker in the case of a nonvalid reference stimulus. It was also interesting to note that this was also differently pronounced in different regions of the brain. For example, the activity of a nonvalid indication stimulus deviated more strongly in the frontal area than in the valid area. In the occipital area, this was the other way round. Here, the response to a valid stimulus was stronger than that of a nonvalid stimulus (cf. Luck et al. 1994, p. 898f.). From this result it can be deduced that attention can act both as an amplifier and as a damper of neuronal activities. It is thus a kind of mediator that achieves focus by shifting neuronal activity, as takes place in a situation in different regions with partially opposite characteristics. In this case, the stimulus activation in the area near the sensor is increased and the intensity in hierarchically higher areas is reduced if a stimulus is valid. The opposite was true for invalids. In the process, attention already anticipated the activity at the sensory level by about 60 to 100 ms (cf. Luck et al. 1994, p. 902).

Attention acts as a mediator in the conflict behaviour of different regions and can help one of these areas to break through by influencing it. The Stroop effect is an example of directed attention and its necessity of energy demand for steering. In a setting, a color is projected as a word onto a screen, but in a different colour than the word is. The automation of word recognition in readers is used, which makes non-reading of a word de facto impossible when it is presented. If the word to be read and the colour presented are incongruent, there is a conflict in perception and processing. This conflict, known as the Stroop Effect, leads to a significant increase in processing time when naming the colour presented, which is not the case when just reading the word. This increase in processing time is also the case, albeit to a lesser extent, if the colour read and the colour presented are the same (cf. Anderson 1996a, p. 94ff.). This result shows that attention can be directed and also steers itself, but this takes time.

The ability to resolve conflicts is also a fundamental characteristic of the brain, which permeates all levels of the system.

3.2.2.5 Conflict Negotiation of Competing Parts

Conflict in the sense of neuronal interaction in this work is understood that two systems, whether they are neurons or neuronal networks, are necessarily connected in an "all-or-nothing" principle underlying competing connection, which allows only one of the conflicting systems to appear victorious. This conflict can lead to contrast formation on the sensory level, as well as to focus on the white team on higher levels and thus to the blanking out of a monkey of the field of vision. The Stroop effect has already shown that the resolution of higher conflicts requires more time.

A review of anatomical, imaging electrophysiological and developmental findings shows that the areas located in the frontal cortex on the rostral–caudal axis are increasingly responsible for more abstract representations and more complex rules with increasing rostral position. The dorsal and lateral adjacent regions are also independently functioning networks, as is the parietal and lateral temporal cortex. This results in a division of labour that makes sense in a changing environment, because the higher processes do not have to change directly with every change in the environment but leave this task at the lower levels (cf. Badre and D'Esposito 2009, conclusions).

The anatomical interconnection also confirms such a structure, because the connection from the rostral to the caudal regions is unidirectional, which could represent a hierarchical arrangement (cf. Badre and D'Esposito 2009, conclusions).

By combining these and other insights, understanding conflict resolution easily leads to an understanding of influential control. Because this model must work and function with the basic principles of neuronal interconnection, the question arises as to how a simple solution can be found in complex situations. To this end, the extended model of the control of behaviour according to Frank and Claus can be applied analogously, which with a simulation of normally distributed evidence values by simple interconnections by means of feedback loops can achieve a contrasting effect and thus a decision, which at the same time experiences an amplification for the future (cf. Frank and Claus 2006, passim). Based on this model, in 2007 and Hyafil created a motive action model, which is negotiated via OFC, LPC, and FPC, respectively. The LPC works as a link between sensory cortex and motor cortex. Therefore, it used to map the contextual stimuli in the LPC, thus giving preference to specific action patterns of these inputs. In parallel, the OFC processes the motivational patterns, which arrive from the paralimbic regions, and lead to preferences within the options (basically the reward expectancy). As LPC and OFC act in exciting interaction with each other, the action patterns strengthened by the strongest motivation are enforced in the LPC. Because, in parallel, the FPC is also excited by the OFC and has the same retroactive effect, a picture emerges of support for the highest motivational pattern despite the negotiation in the LPC. At this point, an asymmetrical connection of the FPC and the LPC leads to an extension of the negotiation cascade. As the inputs of the LPC inhibit in the FPC, the strongest aspects of the LPC are circularly reduced, because the inhibited areas lower the FPC activation in the respective area and therefore reduce the excitation in the OFC. This level is therefore lower again in the LPC, that is, at the origin of the inhibition. At the same time, the connections of the FPC to the LPC excite it, amplifying the less pronounced action patterns from the LPC and OFC negotiation. These inputs generate an amplification of the OFC (the OFC excites the FPC, thus amplifying the same LPC areas), creating a feedback circuit structure that is able to generate and negotiate contrast on the one hand and to keep one of the weaker patterns in a waiting loop on the other. This model is again a general functional principle that can work with simple stimulus–reaction elements. Thus, also the higher levels on the mechanisms of the lower, up to the neurons, produce coherent structures. This finding would also be consistent with the fact that this conflict resolution procedure is similar at all levels and thus explains why some conflicts are conducted up to the anterior prefrontal cortex and others are already resolved in the rostral lateral prefrontal cortex, for example (cf. and Hyafil 2007, p. 596).

A process in which conflicts are likely to be resolved at less extensive levels, even if they are more complex, is expertise. This is the next subject.

3.2.2.6 Expertise

In the course of acquiring skills, that is, achieving expertise, problems are no longer categorised on basis of superficially similar characteristics, but rather on the basis of underlying principles (cf. Anderson 1996a, p. 286). The problem solutions become easier, because the same mechanisms can be applied, which are also correspondingly problem oriented. Here, the already advanced abstraction ability of the experts intervenes in the process in a beneficial way.

As a further element of the development of problem representation and thus as a characteristic of the training of expertise is the creation of a technical language, a technical jargon, with specific meanings. This development economises the thinking and communication skills and thus the understanding as well as the necessary skills for the presence in the (working) memory (cf. Anderson 1996a, p. 287f.).

In an experiment by Chase and Simon in 1973, the economisation of thinking tasks was combined with chunk theory. Thus, it could be clarified what makes the difference between a chess amateur and a chess expert in a memory task concerning the game. Their approach goes back to a test arrangement by de Groot. Both amateurs and experts were shown a chessboard with a playing position for a few seconds, and it was then taken out of view. Now the test persons should adjust the position on a board. It could be shown that beginners could set up about 5 figures, professionals up to 20, correctly, which was done by combining figure positions into a pattern. From this 1965 insight, Chase and Simon developed the idea that it would be interesting to see how this compilation of patterns works. They therefore set the task in a slightly different form: a constellation set up on a test should be reproduced by test persons on a board immediately next to it. In the process, the scientists observed how many figures were set up with one turn of the gaze. This amount was categorised by them as an individual's corresponding "chunk." Experts often processed figure relations as one piece of information, such as a peasant chain. The ability to combine these into a pattern was justified by the fact that chess experts had already seen and processed many more variants of game constellations, which is why they were able to construct a hierarchical model from these, similar to the example given with counting and representation on one hand. Thus, the evidence of chunks as units of meaning of knowledge representation was confirmed (cf. Chase and Simons 1973, abstract).

This experiment was reviewed in 1998 by Simon, together with another scientist, Gobet, and compared with recent studies using other apparatuses. The existence of chunks was again confirmed as a psychological reality (cf. Gobet and Simon 1998, abstract).

If one now speaks of expertise as more comprehensive knowledge, which develops into a more abstract, because hierarchically higher organised, knowledge with corresponding knowledge units, one quickly comes to the topic of intelligence or its measurement by means of IQ tests. Such tests are held under a wide variety of aspects and are certainly criticised, as both the characteristics and the execution of the tests or their comparisons are discussed.

As early as 1932, different emphases showed that IQ tests are not comparable, because these designations place logical thinking or arithmetic in the foreground or consist of several modules (cf. Graf 1932, abstract).

Nevertheless, IQ tests show high stability with regard to the performance of the test persons, even if they are examined over several years or decades. A study conducted in Scotland showed that 50% of the cognitive status of people over 70 years of age could be predicted by the IQ test results obtained at the age of 11 years. The other 50% were influenced by life circumstances, especially fitness, smoking, and education, and were therefore explainable. As this study was conducted by a "mixed methods" approach, factors such as brain damage could be taken into account, and it could also be shown that the thickness of the cortex in old age correlated with IQ in adolescence (cf. Underwood 2014, passim).

An apparently important aspect that affects IQ test performance is time pressure. In some tests, time pressure is a limiting factor. Because the biorhythm has an effect on reaction times, tests designed for this factor may be distorted (cf. O'Connor and Molly 1991, abstract).

However, there may be a connection between intelligence and working memory, and the question arises which of the two influences is which. So, is it the working memory that allows better performance and thus provides better IQ test results, or is it the intelligence that allows the working memory to perform better, which helps to achieve advantages in IQ tests (cf. Krajewski et al. 2008, abstract; Ricken and Fritz 2006, abstract)? However, there is a strong risk of circular reasoning with this question, because a better working memory provides advantages when acquiring cultural techniques, that is, arithmetic, writing, and reading (cf. Vock and Hasselhorn 2010, abstract). However, that is exactly what an intelligence test measures.

These approaches again lead to the question of whether the existing "microworlds" can be dissolved under one merger. At least in the area of expression, the neurosciences show a clear confirmation of neuronal correlates with psychological observations: the idea of space and the solution of linguistic problems require different areas in the brain (cf. Guttmann 2000b, p. 5).

Furthermore, positive emission tomography (PET) studies have shown that subjects with better performance in solving abstract tasks had less activity to conclude, that is, less effort, suggesting that this "is due to very basic processes" (cf. Anderson 1996a, p. 440). In conjunction with the realisation that experts can develop higher abstractions through experience and thus process them more efficiently, it can be concluded that, in connection with the fact that the circular reasoning leads to the conclusion that "intelligence is, what the intelligence test measures," persons with a higher IQ possess a basically more efficient working brain. These people can come to higher abstractions with less expenditure but faster.

However, this finding also shows that expertise leads to better and denser networks in one area of the brain, to integrate more information and more comprehensive content, as well as include more content relationships.

From particularly capable and intelligent persons, the leap is no longer far to the concept of genius. These people, who are often described as very outstanding, are also often characterised as highly creative, which leads to the next topic.

3.2.2.7 Creativity

As already mentioned in Disciplinary connection (2.5, Creativity research), creativity is above all associated with a special achievement. Thus, creativity of scientists is somewhat generally defined in such a way that a creative idea must be new, but also useful or applicable for at least a part of the population (cf. Carson 2010, p. 5).

Andreasen defines creativity similarly as the process of a person creating a product that has both novelty value and utility (cf. Andreasen 2005, p. 17).

In creativity research, there is a distinction between ordinary and extraordinary creativity. Everyone is capable of ordinary creativity. One example of this is free speech, which can produce something new in the new sequence of words while complying with the language-specific rules: this creates something new, mostly useful (cf. Andreasen 2005, p. 63). A further aspect is the ability of human beings to solve problems in everyday life. In this case, functional fixation is an interesting aspect, namely, reducing the material available for problem solving in the immediate vicinity to its everyday use, thus blocking an alternative application that could be a useful option in the situation (cf. Gerrig and Zimbardo 2007, p. 306f.). Overcoming such a blockade by recognizing an alternative function of everyday objects is already a special creative process.

For the ability to be creative, divergence or divergent thinking was seen as a key component. With this background, objective measurements in experimental tests were aimed at and used in the quantification of creative potential in test variations similar to IQ tests. An example of one such is the "Torrence Test of Creative Thinking" (TTCT). However, such tests are generally criticised because they lack predictive power. Nevertheless, they are still widely used in the absence of alternatives (cf. Andreasen 2005, p. 34f.).

These examples show that a distinction between ordinary and extraordinary creativity cannot presumably be made from the processes in general, but only in the depth and peculiarity of the expression. Therefore, a focus of creativity science also lies in the field of process research. In this, a distinction is made between "spontaneous" and "deliberate" creativity (cf. Carson 2010, p. 60); these are subdivided into sub-areas but are essentially similar. The general course of the process is described in reference to the sociologist Graham Wallas, who described it in his book *The Art of Thought*, published in 1926, as divided into several stages (Carson 2010, p. 62):

- Preparation
- Incubation
- Enlightenment/insight/knowledge
- Verification

The essential distinction between the two paths of creativity is that on the spontaneous path, by letting go of thoughts in the incubation phase, a shift of activity into the so-called *default network* takes place (Carson 2010, p. 60). The default network is found in the medial prefrontal cortex, the cuneus, in the posterior cinguli gyrus, and in the lateral parietal cortex (cf. Macedonia and Höhl 2013, p. 60). In contrast to conscious thinking, this is to be found in the upstream levels mentioned in Sect. 3.2.2.5, Conflict negotiation of competing parts. In these conflicting levels, which are usually played off against each other by the decisive "all-or-nothing" principle, something else happens and leads to a decision. An activation of these upstream and hierarchically lower areas promotes more and newer ideas as well as more productive ones, and this in parallel. The step-by-step path, on the other hand, activates only one thought after another and makes use of the frontal and lateral areas (Carson 2010, p. 60); these are the areas of higher control (Badre and D'Esposito 2009, conclusions).

Creativity can be explored, for example, in improvised music playing, as is done in jazz improvisation. A study conducted in 2008 was able to show that improvisation expressed itself through a dissociative pattern in the brain activity recorded by fMRI, which showed increased excitation in the medial prefrontal cortex and reduction in the dorsolateral prefrontal cortex and lateral orbital regions (cf. Limb and Braun 2008, abstract). Another study on rappers showed the same changes in brain activity when these no longer played conventional songs, but spontaneously formed rhymes and rapped themselves. A logical activation of the language areas during the language activation was also confirmed; this was higher in the freestyle variant than in the reproduction, which could not be distinguished in areas of the motor cortex. Consequently, the same motor skills are provided, but the language skills are different. A connectivity study associated with this experiment showed that other parts, including the amygdala and the anterior insula, were involved in these processes: a positive correlation was seen here (cf. Liu et al. 2012, discussion and abstract).

Another aspect of creativity is lateral inhibition, which has an important role in filtering irrelevant incoming information so that the focus can be on the important areas. If this is reduced, more gifted people can achieve higher levels of creativity (cf. Carson et al. 2003, abstract). However, lateral inhibition was also associated with schizophrenia and investigated in experiments. Such experiments showed that visual fine tuning, which is largely based on lateral inhibition, correlates negatively with the development of schizophrenia or pathopsychology (cf. Green et al. 2011, chapter "Visual Tuning and Theories of Pathopsychology and Schizophrenia"). This finding also agrees with the fact that among highly creative people there is an increased probability of mental illness (cf. Andreasen 2005, pp. 95, 138f.).

Because a higher probability of very creative achievements was recognised in more intelligent persons (cf. Carson et al. 2003, p. 499), the question of association with IQ is interesting. However, when linked to this, a correlation was found that existed only up to a threshold value and was then no longer present. This threshold was between 120 and 130 IQ points, which is interesting in this context, because there was one exception, namely scientists, especially physicists, who were exceptionally good at mathematics (cf. Andreasen 2005, p. 30). However, there may be connections with the way in which intelligence is measured. If a distinction is made between fluid and crystalline intelligence, intelligence is found more in the crystalline intelligence than in the fluid intelligence, that is, with direct correlation (cf. Sligh et al. 2005, abstract). There are also criticisms that these differences would be negligible (cf. Kim 2005, abstract), which is why the theory of the limit value could be seen as relativised but is perhaps an indication of a special context.

Lateral inhibition has the effect that increased contrast takes place on different levels. By eliminating[7] these, creative processes in their entirety become easier, because similar valuable states[8] meet in this way.

The topics covered in this chapter are reflected in the following established models and their components. Together, they thus form a basis for the state of the art, which is important for further understanding of the work.

3.3 Models of Cognitive Structures

In this chapter, the models based on the previously mentioned scientific documents are briefly presented to derive the research gap and the research-guiding questions.

3.3.1 ACT-R

The theory of "adaptive control-of-thought rational" (ACT-R) is the theory developed by John R. Anderson and derived from cognitive science.

For Anderson, complex cognition arises from procedural and declarative knowledge, which is represented by "production rules" and "chunks." These are created in large quantities by encoding the environment and are then available for human thought. Activated by a special context, these can then be selected and used (cf. Anderson 1996b, p. 355).

According to the ACT-R theory, the performance of human cognition depends on the amount of stored information and its effective use (literally, "deployment") (cf. Anderson 1996b, p. 355).

During the development of this theory, a computer program was written for verification purposes, which comprised more than 500 production rules, and with which the acquisition and change processes could be statistically reproduced coherently in the interconnection with real investigations (cf. Anderson 1996b, p. 356).

[7] The elimination of lateral inhibition is called cognitive disinhibition and is an expression of reduced inhibition of simultaneously incoming signals, likewise in the processing of sensoric information (cf. Carson 2010, pp. 80–83). Exemplarily, a person driving a vehicle is able to talk to a front-seat passenger until the full attention is shifted to a changing situation, such as a suddenly appearing threat. In this situation the full attention is focused on the new situation but none at the former conversation. This phenomenon, also known as tunnel view, is an extreme example for a reduction of stimuli.

[8] Similarly, valuable in this context means that the resolution potential in conflicting situations at the starting values is less different, which suppresses a feedback loop-related shift towards the option usually possible from higher levels.

3.3.1.1 Presumptions

In this theory, Anderson makes the following presumptions:

Declarative and procedural knowledge are closely linked, because both the states and actions of production rules are defined by declarative knowledge, and new declarative structures can be created by production rules among many other aspects (cf. Anderson 1996b, p. 356).

The declarative knowledge is represented by chunks, which are a schema-like structure consisting of an "isa-pointer"[9] to specify the category and additional references to encode the content (Anderson 1996b, p. 356).

Procedural knowledge is represented by productions: productions are reacted to with rules that follow a certain goal, or with subgoals for a goal. By creating subgoals, human thought implies an abstract and hierarchical structure, which is also expressed in the ACT-R theory in its production system (cf. Anderson 1996b, p. 357).

The sum of the parts are the elements listed here and shown in Fig. 3.3: "These are the parts I have my mind. These parts are the production rules and the chunk structures that represent the long-term knowledge and the evolving understanding of the problem." (Anderson 1996b, p. 357).

3.3.1.2 Knowledge Acquisition

The ACT-R theory is above all a sensory theory whose knowledge structures are derived from the encoding of the environment (cf. Anderson 1996b, p. 359). Therefore, the circular reasoning, which sees declarative structures as generated by procedural processing as well as procedural processes in the chunks, that is, the declarative, as being fundamental (cf. Anderson 1996b, p. 357), is broken up by this approach and sets the beginning to the encoding, albeit still unstructured, of declarative contents.

One aspect assumed in the ACT-R theory is that main features are available as parts of an object before recognition, but only with the focus of attention does an exact identification become possible (cf. Anderson 1996b, p. 359). This assumption was based only on visual processing from the progress made at that time.

Furthermore, Anderson assumes that the processing of visual stimuli by a translation into chunks takes place on a higher level (cf. Anderson 1996b, p. 359).

In ACT-R theory, production rules are created by specifying the change of chunks based on experience. This approach creates a very strong link to the "instruction" that therefore all is learned through imitation (cf. Anderson 1996b, p. 359). This approach is confirmed by various studies which show that abstract, step-by-step instructions are less successful than learning with examples.

[9] "Isa-pointer" is the combination of "is" and "a" and the pointing towards something. It therefore points literally towards the associated meaning or aspect, which expresses that associations linked to information themselves are chunks with their own "isa-pointers."

Nevertheless, abstract instructions also have two important tasks in ACT-R theory. On the one hand, they serve to make the right paths stand out to bridge or capture the change. On the other hand, these instructions can serve to make it easier to find the way through the creation of intermediate goals (cf. Anderson 1996b, p. 359).

What is special about the ACT-R theory is the simplicity of the fact that environmental information is encoded in chunks, whose changes and examples lead to the creation of production rules (cf. Anderson 1996b, p. 359).

3.3.1.3 Knowledge Application

As people collect innumerable amounts of information and thus memories about life, the question arises as to how precisely currently context appropriate to the knowledge can be activated, because with artificial intelligence this leads to an increasing degree of deceleration (cf. Anderson 1996b, p. 360). In ACT-R theory, this problem is solved by a two-pass solution, one based on the probability derived from Bayesian inference and linked to experience, and the other relates general utility to contextual appropriateness, thus activating the most appropriate knowledge content. Anderson expresses this in a simplified formula: Activation-Level = Base-Level + Contextual-Priming (cf. Anderson 1996b, p. 360).

Anderson applied this theory to memory, categorisation, and problem solving, and compared it with results obtained from behavioural observations. As his values obtained from the computer program were consistent, he was able to confirm his theory (cf. Anderson 1996b, pp. 360-364).

3.3.1.4 Summary

In the selection of memory (knowledge/information), in the categorisation of objects or in the selection of a strategy, both test persons and ACT-R react sensitively to previous information as well as to the information about appropriateness in the respective current situation (cf. Anderson 1996b, p. 364). Although people are not aware of this, they are very close to the optimum in their choice, following the Bayesian perspective (cf. Anderson 1996b, p. 364). This Bayesian inference happens unconsciously and often does not match the conscious evaluations of the people. However, it expresses how "we achieve great adaptiveness in knowledge deployment by simple statistical inference" (cf. Anderson 1996b, p. 364).

3.3.2 Integrated Theory of Mind

The "Integrated Theory of Mind" is an applied model based on Anderson's ACT-R theory, which in its extension is based on more recent findings in the cognitive sciences and neurosciences. Anderson has refined his previous approaches to the extent that he has presented them above all as a model for the mode of action of the working memory and related them to other elements of the brain, whereby he represents several modules and was thus able to generate a more coherent cognition through their integration (cf. Anderson et al. 2004, p. 1036).

The working memory was defined as a specially located, higher-level buffer system in the respective processing centres, which has a limited processing capacity and holds the currently activated information ready for a central production system. This central production system does not react to the information stored in the background, but to the information stored in the buffers of the other modules. This central area is able to act on the buffers and demand certain actions from them. The buffers themselves perform similar tasks to what Baddeley calls the working memory. For example, a goal buffer is used to keep the actual task control activated (cf. Anderson et al. 2004, p. 1037).

The buffers are not limited either: they include at least the goal buffer, which is important for target tracking; the retrieval buffer, which is important for the activation of long-term memory and declarative memory content; the perceptual-motor modules, which include the manual buffer; and the visual buffers. with one located in the dorsal "where" path and another in the ventral "what" path, as well as elements of the auditory buffers that are divided into those already mentioned (cf. Anderson et al. 2004, p. 1038).

Important at the central elements lying in the basal ganglia is their production task. These elements are seen in the ACT-R-5.0 theory (an already extended theory of ACT-R) as the place of production rules and thus as part of procedural memory; this describes the function of the striatum as pattern recognition, the pallidum as conflict resolution, and the thalamus as control of the execution of production processes (cf. Anderson et al. 2004, p. 1038).

An essential task of the production rules is updating the buffers in the ACT-R architecture. This action is congruent with the theory of specialised fields and expresses itself in its own critical cycle: the buffers keep active representations determined by the environment and the internal modules, whereupon patterns in the buffers are recognised. The patterns trigger the firing of a production and update the buffers for a new cycle, which should all together take about 50 ms (cf. Anderson et al. 2004, p. 1038).

> "Thus, a production rule in ACT-R corresponds to a specification of a cycle from the cortex, to the basal ganglia, and back again. The conditions of the production rule specify a pattern of activity in the buffers that the rule will match, and the action specifies changes to be made to buffers." (Anderson et al. 2004, p. 1038).

The architecture allows and expects many parallel processes. But there are two bottlenecks: first, each buffer is limited to a declarative unit of knowledge, called chunk, and second, only one production rule can fire per cycle (cf. Anderson et al. 2004, p. 1038).

In the modules, the declarative and procedural memory in particular are briefly addressed here, as these have undergone some important changes.

3.3.2.1 The Declarative Memory Module

Together with the procedural system, it is the core of cognitive performance in the ACT-R model and is based on a set of equations and parameters (cf. Anderson et al. 2004, p. 1042).

A general formula for activity theories is the activity of a "chunk," that is, the sum of the basic activity, which expresses the importance of the chunk in the past, and the reference to actuality brought into the context by the associations (cf. Anderson et al. 2004, p. 1042):

$$A_i = B_i + \sum_j W_j S_{ji}$$

A_i is the activation of the chunk i, which consists of its basic activation B_i, and the sum of W_j and S_{ji} results in W_j that corresponds to the "weighting" of the elements active for the current target; S_{ji} is the strength of the association of the elements j on chunk i. The probability of retrieval depends on this activation, as does its speed (cf. Anderson et al. 2004, p. 1042).

The theory of declarative memory gives a natural description of the explicit-implicit distinction. Explicit content is located in specific chunks, whereas implicit content reflects the sub-symbolic processes that regulate the availability of these chunks (cf. Anderson et al. 2004, p. 1044).

3.3.2.2 The Procedural Memory

The control of the system over cognition as well as its adaptability is achieved through the procedural structures: these are thus at the same time the link between the otherwise seemingly fragmentary elements that would interact with the information contained in the buffers. There are multiple production rules in this model, which compete with each other in the production system. In deciding which of the production rules to apply, Anderson uses various calculations similar to those used for chunks, which include the utility calculation provided with a random variation of noise, the probability of success calculation provided with a damping as well as the topicality reference given by the context (cf. Anderson et al. 2004, p. 1044f.).

All production rules are already defined in these assumptions, which limits their degree of freedom in application. To resolve this, Anderson defined the actual process of the emergence of rules of production in more detail to make the anticipation of the elements to be applied obsolete through the process and thus generate a more general theory (cf. Anderson et al. 2004, p. 1045). The "production compilation" can bring together successful pairs of productions in such a way that the effect of both is produced.[10] This combination allows the old rules to continue as well as the

[10] This is not possible in specific processes, such as in the sensorimotor area; therefore, it has limitations: this is caused by the sequence of processes, which means some have to be finished before another one can be initiated. For example, a hand must be moved towards a glass before grabbing it.

new rule. These rules, as are all others, are in competition for the application, that is, the selection. Because the new rule does not have an initial value when it is created, it is assigned one by combining the initial damping value as a function (subtraction) of the probability value of the old rule. The same action happens analogously with the costs of the production rule. Because the probabilities of the two rules are usually essentially the same, but the costs of the new rule are lower, the new rules become superior (cf. Anderson et al. 2004, p. 1045).

3.3.2.3 General Discussion

ACT-R, similar to some others, is a theory that focuses mainly on the aspect of integration mentioned by Newell in 1973. However, these theories and approaches are still far from providing an all-encompassing solution for cognition and its processes. What is special, however, is the attempt to make a prediction through the process-oriented approach and not the "after-statement." Criticisms that this model artificially approximates reality by using fixed values are not correct but could best be served by simply removing the parameter estimate (cf. Anderson et al. 2004, p. 1057). A positive aspect of the integrative approaches is the openness to further methodological possibilities such as the fMRI.

The whole theory is not yet complete and has not yet integrated some modules of the brain. However, it is not correct that all neuronal processes pass through the basal ganglia because there are direct connections of brain regions; this would create opportunities to circumvent the production areas of the ACT-R. Nevertheless, it can be stated that the existing consensus in cognitive science of parallel processing in the specified areas is being integrated by this theory. In doing so, the corresponding modules maintain and pursue the goals, keep their own history, and understand the world around us through their interaction patterns. This information is coordinated via buffer systems and a central production area and subsequently translated into action. The sub-symbolic learning and achievement mechanisms in ACT-R make these actions appropriate, so that the elements appropriate in each situation become active (cf. Anderson et al. 2004).

Thus, a certain fundamental truth might be inherent in these elements (cf. Anderson et al. 2004, p. 1058).

3.3.3 Four Levels of Personality

The model of the four levels of personality of Gerhard Roth and Nicole Strüber encompasses the unconscious as well as the conscious self. At the level of the unconscious self they see the lower and middle limbic system, which is primarily responsible for vegetative-affective behaviour and at the middle level for emotional conditioning, reward, and motivation (cf. Roth and Strüber 2014, p. 63f.) as well as for unconscious behaviour evaluation, emotion formation, and control (cf. Roth and

Strüber 2014, p. 68). On the conscious side there is the upper limbic level, which is important for conscious feelings and motives as well as socialisation and education (cf. Roth and Strüber 2014, p. 83). As the last and also conscious level they see the cognitive linguistic area of the brain, which is responsible for the sensory areas, motoric areas, and their control, cognitive-associative areas with their complex meaningful perception, memory, and language, as well as the executive areas of behavioural planning and preparation (cf. Roth and Strüber 2014, p. 87).

This model is important because it expresses a mode of action confirmed by connectivity studies from the limbic system into the cognitively conscious realm, which is not diminishing but can only lead to changes in the amygdala by detours and changes in the release of neurotransmitters such as oxytocin and by neurogenesis in the hippocampus, which changes in the long term.[11] This explanatory approach is indirectly coherent[12] through psychotherapeutic process research and thus shows the influence of unconscious processes and emotions on cognition.

3.3.4 Hierarchical Control Model

In Sect. 3.2.2.5, it was mentioned that the processes existing in the neuronal process can compete. This is the case in parallel processes in different specialised areas as well as on the cell level, where neighbouring neurons influence each other directly or indirectly. Through this parallelism, processes or decision processes become necessary, which dissolve them and thus execute, on all levels of the neuronal system, the "immanent" "all-or-nothing" principle. In the model of hierarchical control, this is achieved in a self-organised and unspecific negotiation process through exciting and inhibitory feedback (cf. Rösler 2011, p. 328f.).

This model is based on a cascade model presented by Koechlin in 2007, which divides the frontal cortex into hierarchically ascending areas. The further they extend forward and outward, the higher. This aspect was confirmed by Koechlin together with D'Esposito in a complex four-stage reaction selection experiment, which included extensive dependencies of stimulus configurations and was examined and evaluated by fMRI (cf. Badre and D'Esposito 2007 in Rösler 2011, pp. 323–328).

The model of hierarchical control shows that two different states of information at the same level, such as two actions, are in competition with each other as an option for an external stimulus for the implementation. For this conflict to be resolved, the decision-making instance on the next higher level is needed. The latter makes the decision by the fact that the two lower states of information enter into a kind of activation competition, because they both have a stimulating effect on the

[11] Cf. Roth and Strüber 2014, Chapter 7: Psychological disease and personality traits.

[12] Cf. Roth and Strüber 2014, Chapter 8: Psychotherapy.

decision-making authority and at the same time inhibit the competing network. One of the two subsystems then prevails and becomes "activated" (Rösler 2011, p. 338).

An important aspect here is that the decisive system does not have to have any specific representations of the subsystems,[13] but works with nonspecific information through feedback (cf. Rösler 2011, p. 329).

A further point is that the second data layer/activity pattern is not lost, but only steps into the background. As a result, it would be easier to retrieve one of these, because it is only on hold (Rösler 2011, p. 342).

That there are often several options in life is taken into account by the fact that this system is represented over several levels and can thus be implemented as a cascade model (Rösler 2011, pp. 342, 329).

As already mentioned in Sect. 3.2.1, Visual perception with lateral inhibition, this model goes back to the lowest levels, or, vice versa, it is used continuously from these to the highest levels. The question now arises what happens in the brain when it acts in free mode, that is, these hierarchical states are less active.

3.3.5 REST State

The "random episodic silent thought" state (REST state) has been investigated by Andreasen in a PET study. She was interested in the phase of rest, used as a control state in other investigations, in which, in her opinion, free association takes place and not activity. For them, free association is the process that is achieved by the free wandering of thoughts with the elimination of motor or sensory input. This mental process creates the connection of nonassociated areas without conscious performance and thus leads to creativity (cf. Andreasen 2005, pp. 72, in combinatio with 70).

An important basis for this is the distinction between two elements of declarative memory, which have already been briefly hinted at in Sect. 3.2.2.2, Memory systems. In semantic memory, generalised content, that is, content that has already been abstracted and related, is depicted, whereas in episodic memory, personal experiences are referenced in relation to time and space, thus enabling, among other things, the ability to introspect. The episodic is associated with free association, which is why it was used as the state to be investigated for the PET study (cf. Andreasen 2005, p. 71).

To investigate the resting phase as a phase of free association, Andreasen used in her study a dull and boring semantic memorising task as a control state, which had to be completed by the test persons. As experimental states, she used a simple, "focused episodic memory" (FEM) about the previous daily routine and a

[13] This also works in sense of specialisation, because otherwise specific representations would need to be available more often than once. Surely some information patterns overlap as well as different associations work together, but despite that, these are not stored and mapped with all properties in all specialised systems.

subsequent phase of rest that reflected the REST state. In the REST state and in FEM, two areas overlapped and were active. On the one hand, the associative areas of the inferior frontal lobe were responsible for social attention and value feeling. On the other hand, parietal associative regions were involved in both states. However, an increased activity only in the REST state could be observed in the associative cortex[14] responsible for the integration of information from different brain areas, which in sum with the areas also active together in the FEM shows that in the rest phase activity prevails above all in the associative areas (cf. Andreasen 2005, p. 73).

Thus, an essential part of this state is that a freer association and thus a freer integration of experience and information, etc., becomes possible through the omission of the focusing areas, and thus new ideas can be created.

3.3.6 CREATES Model of Creativity

In this approach conceived by Shelley Carson, a link is established between the creative process and findings from the neurosciences, thus providing a theoretical model for the neural activity patterns and states required in the various process stages.

This model is based on three principles (cf. Carson 2010, p. 14f.):

- Creatively productive individuals reach alternative states in the brain.
- Creatively productive individuals change states in the brain depending on the task.
- The change between the states can be trained.

She approaches modeling from the side of the creative process to name the brain areas and different states involved in the respective phases of the process, which overlap, that is, are not discrete. The whole concept is described under *Connect, Reason, Envision, Absorb, Transform, Evaluate, and Stream* (CREATES) (Carson 2010, p. 15).

Connect is a state of non-focusing that enables ideas to begin to flow and divergent thinking to develop (cf. Carson 2010, p. 123). Thus, the characteristics of this state are divergent thinking, unusual associations, intrinsic motivation, the resting of the censor, the response to novelty and complexity, and the reduction of judgement (cf. Carson 2010, p. 124). Neuronally, this is expressed by the fact that the alpha waves in the prefrontal lobe increase to reduce attention, but are still present for active searching. Furthermore, the right hemisphere is relatively more activated than the left hemisphere, which is associated with the distance processing of semantic

[14] The associative cortex links different cortices by using different associations circuits (within areas on one hemisphere), commissure systems (linking the two hemispheres), and projecting systems (linking the deeper lying sensor and motor brain areas as well as the spinal cord) (cf. Gramann and Schandry 2009, p. 36f.).

processes. In the left hemisphere, the semantic area and the association centres are activated (cf. Carson 2010, p. 141f.).

For Carson, reason is converging thinking that is conscious, focused, and sequential (cf. Carson 2010, p. 155). This state is described by the properties of conscious and intentional control of thought processes, a reference to reality or practicability, and a sequential mode of operation (cf. Carson 2010, p. 160). The activation of this activity state leads to an increased activation of the executive centres in the prefrontal cortex as well as the association centres of the left hemisphere. Alpha and theta waves decrease in the frontal areas (cf. Carson 2010, p. 165).

In the envision configuration (insight), states of mental visualisation are used to enable metaphorical relationships and transformations with which different concepts can be connected. The prefrontal cortex creates a connection between abstract thinking and imagination as well as with storage areas, so that memories and mental images are possible as the essence of insight (cf. Carson 2010, p. 105). In this way, novelty is preferred and cognitive disinhibition is created. The connection of mental images with hypothetical thinking is another characteristic of this state (cf. Carson 2010, p. 107). This state manifests itself in an increased activation of the association centres as well as a partial activation of the executive areas, because these are partly required for the maintenance of created images (cf. Carson 2010, p. 113). Therefore, when creating a new (design) object, the right prefrontal cortex as well as the left superior parietal cortex and the left associative centres are active, whereas the right parietal cortex is relatively deactivated (cf. Carson 2010, p. 114).

In the absorb state one opens oneself mentally to new ideas and experiences, which are absorbed above all without evaluation, which fascinate and attract attention (cf. Carson 2010, p. 74). This state enables the so-called *opportunistic association* (cf. Carson 2010, p. 75), which creates no evaluation, or categorisation, but only a purely free exploration (cf. Carson 2010, p. 78), and makes this possible, among other aspects, with cognitive disinhibition (cf. Carson 2010, p. 81). The goal of reducing the active filters leads to a relatively higher activation in the right hemisphere and a general shift of EEG waves into the alpha and theta band, which is associated with a relaxed receptive state (cf. Carson 2010, p. 84).

In the transform phase, an activation takes place in which one is conscious of oneself and dissatisfied or burdened with oneself, and thereby may transform emotional experience into creative works (cf. Carson 2010, p. 208). Thus, in this state one is self-centered, slightly negative, and unsatisfied (cf. Carson 2010, p. 211). This activation expresses itself in an increased activity of the loops of the limbic system and the emotional loops, thus the "default network." The ego area and the executive areas are also active. In addition, the *behavioural inhibition system* causes a retreat from the environment (cf. Carson 2010, p. 221).

In the evaluation state, the results of the work are subjected to a critical examination, which requires active assessment, focused attention, and impersonality (cf. Carson 2010, p. 183). The necessity of knowing about reference norms in the respective specific environment is just as important here (cf. Carson 2010, p. 187) as the ability to reflect on personally distorting influences (cf. Carson 2010, p. 189f.). In addition to the prefrontal cortex, the anterior cingulate cortex and the orbitofrontal cortex, which are jointly responsible for categorisation and assessment, are to be mentioned as active areas for this condition (cf. Carson 2010, p. 197).

In the stream activity pattern, all abilities and skills flow for the improvised solution of a problem. This condition is more commonly known under the term established by Csíkszentmihályi, "flow" (cf. Carson 2010, p. 234). Time becomes blurred, just as self-perception fades into the background, and the focus is totally on the task at hand. A certain level of expertise is necessary and leads to intrinsic motivation and action with trained impulsiveness, that is, improvisation (cf. Carson 2010, p. 236f.). Neuronally, this state is expressed in a relative deactivation of the left executive areas as well as an increase of the premotor areas, the associative areas of the temporal lobe, and a slight activation of the reward centre (cf. Carson 2010, p. 246).

These states are a theoretical and unconfirmed derivation from various experiments that Carson has associated with the process of creativity. It refers to the fact that creativity expresses itself above all in different activity patterns and not in different kinds of connectivity between persons (cf. Carson 2010, p. 263): this is also consistent with the previous findings of the current state of science. An important approach for the creation of this model was therefore the question of how to approach these activity patterns. Because such activity patterns can be influenced both by learning and by transmitter fluctuations (cf. Carson 2010, p. 263), she has combined this insight with exercises and tests which should make it possible to switch more easily and above all consciously between these brain states to create a more fulfilling access to creativity and thus to options in life (cf. Carson 2010, pp. 261–265).

In summary, it can be said that different brain states are beneficial depending on the state of the process. Achieving this can be learned and practiced. For the phases of creative work, those parts of the networked interconnection that express themselves in a cognitive disinhibition, which mean an increased activity of the associative areas and a reduced activity of the executive areas, are particularly important. According to Arne Dietrich, a special case might be a state known under the term of the transient hypofrontality hypothesis, which by the change in the frontal region with an encapsulation allows an improved processing and isolated change of ideas, which is expressed in the interpretation of the change of the EEG there on alpha and theta waves.

3.3.7 The Voxel Approach

This approach is actually not a specific model, but a special type of evaluation. However, as it is an important basis for the model of thought for this work, it should be briefly mentioned here.

A voxel is a composite abbreviation of the words *volume* and *pixel*, created by Haxby and colleagues in 2001 to evaluate a recognition task taking place under fMRI recording. The test persons were each shown different images of a category and had to recognise occurring image repetitions in these, whereby an occurring image repetition was always associated with a change of perspective. Through the formation of voxels, de facto vectors, of fMRI images, in his case vectors with 5000 components, Haxby was able to calculate similarity by correlation (cf. Haxby et al. 2001). He was able to show that the elements in an object category had a high correlation, whereas those from different categories had a low correlation (Rösler 2011, p. 185f.). This not surprising insight is interesting in this work from another aspect: it shows that the total activation of the brain can be represented as a vector. Thus, the brain as well as its subsystems can also be represented vectorially. Finally, it can be observed that similar contents produce similar vectors, that is, similar activation patterns, both in the entrance and in the internal recapitulation, memory, or imagination.[15]

References

Books and Journals

Anderson, J. R. (1996a). *Kognitive Psychologie. Eine Einführung.* (2nd ed.). Heidelberg: Spectrum Academic Publishing House Ltd.

Anderson, J. R. (1996b). ACT. A simple theory of complex cognition. *American Psychologist, 51*(4), 355–365.

Anderson, J. R., Bothell, D., Byrne, M. D., Douglass, S., Lebiere, C., & Qin, Y. (2004). An integrated theory of the mind. *Psychological review, 111*(4), 1036-1060.

Andreasen, N. C. (2005). *The creating brain. The neuroscience of genius.* New York/Washington, DC: Dana Press.

Baddeley, A. (2003). Working memory: looking back and looking forward. *Nature Reviews Neuroscience, 4*, 829–839.

Badre, D., & D'Esposito, M. (2009). Is the rostro-caudal axis of the frontal lobe hierarchical? *Nature Reviews Neuroscience, 10*, 659–669.

Burton, H., Snyder, A. Z., Conturo, T. E., Akbudak, E., Ollinger, J. M., & Raichle, M. E. (2002). Adaptive changes in early and late blind: a fMRI study of Braille reading. *Journal of Neuroscience, 87*(1), 589–607.

Carletti, B., & Rossi, F. (2008). Neurogenesis in the cerebellum. *The Neuroscientist, 14*(1), 91–100.

Carson, S. (2010). *Your creative brain.* San Francisco: Jossey-Bass, A Wiley Imprint.

Carson, S. H., Peterson, J. B., & Higgins, D. M. (2003). Decreased latent inhibition is associated with increased creative achievement in high-functioning individuals. *Journal of Personality and Social Psychology, 85*(3), 499–506.

Chase, W. G., & Simons, H. A. (1973). The mind's eye in chess. *Visual Information Processing, 4*, 215–281.

Chatterjee, R. (2015). Out of the darkness. Defying expectations, cataract surgery in Indian children in Indian children is endowing them with vision-and shedding light on how the brain learns how to see. *Science, 350*(6259), 372–375.

Cook, I. A., Abrams, M., & Leuchter, A. F. (2016). Trigeminal nerve stimulation for comorbid posttraumatic stress disorder and major depressive disorder. *Neuromodulation, 19*(3), 299–305.

[15] This last aspect derives from the fact that the experiences made before going to bed show the same patterns as the retrieval of this situation during sleep, but chronologically accelerated (cf. Ji and Wilson 2007, passim).

De Bellis, M. D., Clark, D. B., Beers, S. R., Soloff, P. H., Boring, A. M., Hall, J., ... & Keshavan, M. S. (2000). Hippocampal volume in adolescent-onset alcohol use disorders. *American Journal of Psychiatry, 157*(5), 737–744.
Dejana, E., & Betsholtz, C. (2016). Oligodendrocytes follow blood vessel trails in the brain. *Science, 351*(6271), 341–342.
Eysenck, M. W. (2009). *Fundamentals of psychology*. East Sussex-New York: Psychology Press.
Ferris, L. T., Williams, J. S., & Shen, C.-L. (2007). The effect of acute exercise on serum brain-derived neurotrophic factor levels and cognitive function. *Medicine and Science in Sports and Exercise, 39*(4), 728–734.
Frank, M. J., & Claus, E. D. (2006). Anatomy of decision: striato-orbitofronal interactions in reinforcement learning, decision making and reversal. *Psychological Review, 113*(2), 300–326.
Gerrig, R. J. & Zimbardo, P. G. (2007). *Psychology* (18th updated edition). Munich: Pearson Education Germany GmbH.
Gobet, F., & Simon, H. A. (1998). Expert chess memory: revisiting the chunking hypothesis. *Memory, 6*(3), 225–255.
Godden, D. R., & Baddeley, A. D. (1975). *Context dependent memory in two natural environments: on land and underwater*. Stirling: University of Stirling. Department of Psychology. Revised manuscript.
Görz, G., Rollinger, C.-R., & Schneeberger, J. (2003). *Handbuch der künstlichen Ingelligenz*. (4th corrected edition). Munich Vienna: Oldenbuurg Wissenschaftsverlag.
Graf, O. (1932). Zur Frage der Konstanz bei Intelligenzprüfungen nach verschiedenen Prüfungssystemen. *Psychotechnische Zeitschrift*, 7, 84–88.
Gramann, K., & Schandry, R. (2009). *Psychophysiologie – körperliche Indikatoren psychischen Geschehens*. Weinheim-Basel: Beltz Publishers.
Green, M. F., Lee, J., Wynn, J. K., & Mathis, K. I. (2011). Visual masking in schizophrenia: overview and theoretical implications. *Schizophrenia Bulletin, 37*(4), 700–708.
Guttmann, G. (2000b). Ludwig von Bertalanffys Systemtheorie und die „Wirklichkeit" psychologischer Konzepte. In K. Edlinger, W. Feigl, & G. Fleck (Eds.), *Systemtheoretische Perspektiven*. Frankfurt am Main: Peter Lang. Manuscript of the author.
Hamm, S., Junglas, K., & Bredenkamp, J. (2004). Die Zentrale Exekutive als präartikulatorische Kontrollinstanz. *Zeitschrift für Psychologie*, 212(2), 66–75.
Haxby, J. V., Gobbini, M. I., Furey, M. L., Ishai, A., Schouten, J. L., & Pietrini, P. (2001). Distributed and overlapping representations of faces and objects in ventral temporal cortex. *Science, 293*(5539), 2425–2430.
Hebb, D. (2009 [1949]). *The Organization of behavior: A neuropsychological theory*. Mahwah: Lawrence Erlbaum Associates Inc..
Hills, T. T., Todd, P. M., & Goldstone, R. L. (2010). The central executive as a search process: Priming exploration and exploitation across domains. *Journal of Experimental Psychology: General, 139*(4), 590–609.
Ji, D., & Wilson, M. A. (2007). Coordinated memory replay in the visual cortex and hippocampus during sleep. *Nature Neuroscience, 10*(1), 100–107.
Labov, W. (1973): The boundaries of words and their meanings. In: Bailey, C.-J.N. & Shuy, R.W. (Hg.): New ways of analyzing variation in English. Washington, DC: Georgetown University Press. 340-373.
Kahle, W. & Frotscher, M. (2005). *Nervensysteme und Sinnesorgane*. In: Taschenatlas Anatomie. 3rd band (9th revised edition). Stuttgart. Thieme Publishers.
Kempermann, G., Chesler, E. J., Lu, L., Williams, R. W., & Gage, F. H. (2006). Natural variation and genetic covariance in adult hippocampal neurogenesis. *Proceedings of the National Academy of Sciences, 103*(3), 780–785.
Kim, K. H. (2005). Can only intelligent people be creative? A meta-analysis. *Journal of Advanced Academics, 16*(2–3), 57–66.
Kimbrel, N. A., Morissette, S. B., Meyer, E. C., Chrestman, R., Jamroz, R., Silvia, P. J., ... & Young, K. A. (2015). Effect of the 5-HTTLPR polymorphism on posttraumatic stress disorder, depression, anxiety, and quality of life among Iraq and Afghanistan veterans. *Anxiety, Stress and Coping, 28*(4), 456–466.

Kitamura, T., Saitoh, Y., Takashima, N., Murayama, A., Niibori, Y., Ageta, H., ... & Inokuchi, K. (2009). Adult neurogenesis modulates the hippocampus-dependent period of associative fear memory. *Cell, 139*(4), 814–827.

Koechlin, E., & Hyafil, A. (2007). Anterior prefrontal function and the limits of human decision-making. *Science, 318*(5850), 594–598.

Krajewski, K., Schneider, W., & Nieding, G. (2008). Zur Bedeutung von Arbeitsgedächtnis, Intelligenz, phonologischer Bewusstheit und früher Mengen-Zahlen-Kompetenz beim Übergang Kindergarten in Grundschule. *Psychologie in Erziehung und Unterricht*, 55(2), 100–113.

Lee, J., Duan, W., & Mattson, M. P. (2002). Evidence that brain-derived neurotrophic factor is required for basal neurogenesis and mediates, in part, the enhancement of neurogenesis by dietary restriction in the hippocampus of adult mice. *Journal of Neurochemistry, 82*(6), 1367–1375.

Lerner, Y., Hendler, T., Ben-Bashat, D., Harel, M., & Malach, R. (2001). A hierarchical axis of object processing stages in the human visual cortex. *Cerebral Cortex, 11*(4), 287–297.

Luck, S. J., Hillyard, S. A., Mouloua, M., Woldorff, M. G., Clark, V. P., & Hawkins, H. L. (1994). Effects of spatial cuing on luminance detectability: psychophysical and electrophysiological evidence for early selection. *Journal of Experimental Psychology: Human Perception and Performance, 20*(4), 887–904.

Macedonia, M., & Höhl, S. (2012). *Gehirn für Einsteiger* (2nd ed.). Linz/Heidelberg: Self-published.

Macedonia, M., & Höhl, S. (2013). *Gehirn für Fortgeschrittene.* Linz: SchEz -School and Education Centre Linz.

Maguire, E. A., Woollett, K., & Spiers, H. J. (2006). London taxi drivers and bus drivers: A structural MRI and neuropsychological analysis. *Hippocampus, 16*, 1091–1101.

Moore, T., & Armstrong, K. M. (2003). Selective gating of visual signals by microstimulation of frontal cortex. *Nature, 421*, 370–373.

Morey, R. A., Gold, A. L., LaBar, K. S., Beall, S. K., Brown, V. M., Haswell, C. C., ... & McCarthy, G. (2012). Amygdala volume changes in posttraumatic stress disorder in a large case-controlled veterans group. *Archives of General Psychiatry, 69*(11), 1169–1178.

Muir, D. W., Blasdel, G. G., Mitchell, D. E., & Pettigrew, J. D. (1977). A physiological and behavioural study in cats of the effect of early visual experience with contours of a single orientation. *Journal of Physiology, 265*, 615–636.

O'Connor, B. P., & Molly, K. (1991). A test of the intellectual cycle of the popular biorhythm theory. *The Journal of Psychology: Interdisciplinary and Applied, 125*(3), 291–299.

Pavlov, I. P. (2010 [1927]). Conditioned reflexes: An investigation of the physiological activity of the cerebral cortex. *Annals of Neurosciences, 17*(3), 136–141.

Pettigrew, J. D. (1974). The effect of visual experience on the development of stimulus specificity by kitten cortical neurones. *Journal of Physiology, 237*, 49–74.

Ricken, G., & Fritz, A. (2006). Arbeitsgedächtnisleistung bei unterschiedlich guten Rechnern im Kindergartenalter. *Psychologie in Erziehung und Unterricht*, 53(4), 263–274.

Rösler, F. (2011). *Psychophysiologie der Kognition. Eine Einführung in die Kognitive Neurowissenschaft.* Heidelberg: Spectrum Academic Publishing House.

Roth, G., & Strüber, N. (2014). *Wie das Gehirn die Seele macht* (5th ed.). Stuttgart: Klett-Cotta Publisher.

Schandry, R. (2011). *Biologische Psychologie* (3rd completely revised edition). Weinheim-Basel: Beltz Publishers.

Shors, T. J. (2009). Saving new brain cells. *Scientific American, 300*(3), 47–54.

Silbernagl, S. & Despopoulos, A. (2003). *Taschenatlas der Physiologie* (6th corrected edition). Stuttgart: Thieme Publishers.

Simons, D. J., & Chabris, C. F. (1999). Gorillas in our midst: sustained inattentional blindness for dynamic events. *Perception, 28*(9), 1059–1074.

Sligh, A. C., Conners, F. C., & Roskos-Ewoldsen, B. (2005). Relation of creativity to fluid and crystallized intelligence. *The Journal of Creative Behavior, 39*(2), 123–136.

Smith, S. M., & Vela, E. (2001). Environmental context-dependent memory: a review and meta-analysis. *Psychonomic Bulletin & Review, 8*(2), 202–220.

Spitzer, M. (2012). *Digitale Demenz.* Munich: Droemer Publishers.

Stankiewicz, A. M., Swiergiel, A. H., & Lisowski, P. (2013). Epigenetics of stress adaptations in the brain. *Brain Research Bulletin, 98*, 76–92.

Tsai, H. H., Niu, J., Munji, R., Davalos, D., Chang, J., Zhang, H., ... & Fancy, S. P. (2016). Oligodendrocyte precursors migrate along vasculature in the developing nervous system. *Science, 351*(6271), 379–383.

Underwood, E. (2014). Starting young – decades old IQ test records from Scottish children have opened a unique window on how the brain ages. *Science, 346*(6209), 568–571.

von der Malsburg, C. (1973). Self-organization of orientation sensitive cells in the striate cortex. *Cybernetics, 14*(2), 85–100.

Vock, M., & Hasselhorn, M. (2010). Diagnostik der Funktionstüchtigkeit des Arbeitsgedächtnisses bei Hochbegabten. In F. Preckel, W. Schneider, & H. Holling (Eds.), *Diagnostik von Hochbegabung* (pp. 119–139). Göttingen: Hogrefe.

Xie, P., Kranzler, H. R., Poling, J., Stein, M. B., Anton, R. F., Brady, K., ... & Gelernter, J. (2009). Interactive effect of stressful life events and the serotonin transporter 5-HTTLPR genotype on posttraumatic stress disorder diagnosis in 2 independent populations. *Archives of General Psychiatry, 66*(11), 1201–1209.

Online Resources

Bartol Jr, T. M., Bromer, C., Kinney, J., Chirillo, M. A., Bourne, J. N., Harris, K. M., & Sejnowski, T. J. (2015). Nanoconnectomic upper bound on the variability of synaptic plasticity. *ELife, 4*, e10778.

Kempermann, G., Fabel, K., Ehninger, D., Babu, H., Leal-Galicia, P., Garthe, A., & Wolf, S. (2010). Why and how physical activity promotes experience-induced brain plasticity. *Frontiers in Neuroscience, 4*, 189. https://doi.org/10.3389/fnins.2010.00189.

Limb, C. J., & Braun, A. R. (2008). Neural substrates of spontaneous musical performance: an fMRI study of Jazz improvisation. *PLoS One, 3*(2), e1679.

Liu, S., Chow, H. M., Xu, Y., Erkkinen, M. G., Swett, K. E., Eagle, M. W., ... & Braun, A. R. (2012). Neural correlates of lyrical improvisation: an fMRI study of freestyle rap. *Scientific Reports, 2*, 834.

Manning, J. R. (2016). A neural signature of contextually mediated intentional forgetting. *Psychonomic Bulletin & Review, 2016*, 1–31.

Rash, B. G., Ackman, J. B., & Rakic, P. (2016). Bidirectional radial Ca^{2+} activity regulates neurogenesis and migration during early cortical column formation. *Science Advanced, 2*, 1–10.

Zsynth (2006). *Figure lateral inhibition*. From the German Wikipedia, CC BY-SA 3.0 [vom 27 07 16] https://commons.wikimedia.org/w/index.php?curid=1645056.

Other

Seifert, G. (2009). *Langzeitpotenzierung in der CA1-Region – Verschiedene Modulatoren.* Inaugural dissertation. Freiburg: Medical Faculty of the Albert-Ludwigs-University.

Wiesner, C. D. (2004). *Was ist zentral an der zentralen Exekutive? Neuronale Korrelate des vorhersagbaren versus unvorhersagbaren kognitiven Verzweigens in verbal-phonologischen versus figural-räumlichen Arbeitsgedächtnisaufgaben. Eine Studie mit funktioneller Magnetresonanztomographie.* Dissertation. Kiel: Christian Albrecht University.

Chapter 4
Research Gap

Contents

As explained in Chap. 3, "State of Research," there is already much knowledge about the functioning of the brain. This knowledge also includes the field of anatomical backgrounds, from which, in addition to the functioning of neurons, the smallest cells and building blocks of the brain, and their limitation in potential control, their adaptability, is also known, through which learning and neuronal plasticity can presumably be explained (Silbernagl and Despopoulos 2003; Kahle and Frotscher 2005; Seifert 2009).

In the neurosciences, this knowledge is combined into models that offer an interconnection structure through which the ability to organise oneself begins to be explained. One of the findings derived from this is through self-referentiality, the explainable redundancy and decision cascades understandable with this, which are characterised by a nonspecification and thus express the hierarchy in the brain both in anatomical connectivity and in experimentally confirming experimental arrangements (Rösler 2011; Chatterjee 2015; Badre and D'Esposito 2009).

This knowledge is further supplemented by establishing a reference to the cognitive sciences, which applies in particular to cognitive psychology, which draws very similar conclusions from experimental designs in the cascade formations derived from the linguistic field and thus shows a deepening, or hierarchical, specialisation both in information processing and in its retrieval. This concept also has a strong contextual dependency, that is, it is referenced to the respective environment or setting (Anderson 1996a, 1996b; Anderson et al. 2004; Godden and Baddeley 1975; Cook et al. 2016).

This area is rounded off by the creativity sciences, which concern both the measurement of creative achievements and their process research. Through the initial approaches of findings from the connection with imaging procedures in the

K. Testor, *Cognitive Theoretical Foundations of Creativity*,
https://doi.org/10.1007/978-3-030-61736-3_4

neurosciences, one approaches here the knowledge of which areas of the brain are involved in these processes (Andreasen 2005; Carson 2010; Limb and Braun 2008; Liu et al. 2012).

Some models and approaches to creativity and thinking are discussed. In these, ACT-R as well as the "Integrated Theory of Mind" focus in particular on how the brain can align itself in the integration of existing and stored knowledge about the use of working memory into an organ that guides action and responds to its specialisation in a way which is adapted to the situation. However, it is not explained why or how something new can be created.

The four levels of personality divide the brain into conscious and unconscious areas, which focus primarily on the influence of unconscious processes that can be both emotional and affective. Furthermore, in this model close relationships are established to the effect of neurotransmitters and neuropeptides, which show a change occurring on the cellular level, as well as approaches from the genetic predisposition are indicated. Nevertheless, apart from the change supported by neurogenesis, which seems to support aspects of the rigidity of action to break through, no real approaches to creating something new are offered here either.

The model of hierarchical control is an approach that can explain the understanding of decision making through interconnection structures and their self-connection, reminiscent of statistical comparisons. Thus, it does not need a specification in the representation at higher levels, but is able to achieve this through a simple "negotiation routine." But this topic also considers the aspect that shows how already existing knowledge is negotiated and not how something new could emerge from it.

The REST state indicates that there are special neuronal activity patterns that lead to the understanding of creativity and the often-associated patterns of free flow of thought in every human being. This approach shows that creativity can be represented in neuronal activity and can therefore be explained by the findings from the sciences working with it. However, this model does not take this step.

The CREATES model is based on the creative process, which assumes that different activity patterns are important for different phases of creativity. To this end, findings from neurofeedback, which make such activity patterns easier to achieve through training, and neurosciences are incorporated to serve as a handbook for increasing creativity. Nevertheless, it is not explained here how this new concept can arise out of thinking.

Overall, it can therefore be deduced that the question of how thinking on the basis of its functionality or neuroanatomical foundations can lead to the emergence of something new and thus to creativity has not yet been examined in detail, even though many studies have already worked on it in peripheral areas.

References

Books and Journals

Anderson, J. R. (1996a). *Kognitive Psychologie. Eine Einführung.* (2nd ed.). Heidelberg: Spectrum Academic Publishing House Ltd.

Anderson, J. R. (1996b). ACT. A simple theory of complex cognition. *American Psychologist, 51*(4), 355–365.

Anderson, J. R., Bothell, D., Byrne, M. D., Douglass, S., Lebiere, C., & Qin, Y. (2004). An integrated theory of the mind. *Psychological Review, 111*(4), 1036–1060.

Andreasen, N. C. (2005). *The creating brain. The neuroscience of genius.* New York/Washington, DC: Dana Press.

Badre, D., & D'Esposito, M. (2009). Is the rostro-caudal axis of the frontal lobe hierarchical? *Nature Reviews Neuroscience, 10*, 659–669.

Carson, S. (2010). *Your creative brain.* San Francisco: Jossey-Bass, A Wiley Imprint.

Chatterjee, R. (2015). Out of the darkness. Defying expectations, cataract surgery in Indian children in Indian children is endowing them with vision-and shedding light on how the brain learns how to see. *Science, 350*(6259), 372–375.

Cook, I. A., Abrams, M., & Leuchter, A. F. (2016). Trigeminal nerve stimulation for comorbid posttraumatic stress disorder and major depressive disorder. *Neuromodulation, 19*(3), 299–305.

Godden, D. R., & Baddeley, A. D. (1975). *Context dependent memory in two natural environments: on land and underwater.* Stirling: University of Stirling. Department of Psychology. Revised manuscript.

Kahle, W. & Frotscher, M. (2005). *Nervensysteme und Sinnesorgane.* In: Taschenatlas Anatomie. 3rd band (9th revised edition). Stuttgart. Thieme Publishers.

Rösler, F. (2011). *Psychophysiologie der Kognition. Eine Einführung in die Kognitive Neurowissenschaft.* Heidelberg: Spectrum Academic Publishing House.

Silbernagl, S. & Despopoulos, A. (2003). *Taschenatlas der Physiologie* (6th corrected edition). Stuttgart: Thieme Publishers.

Online Resources

Limb, C. J., & Braun, A. R. (2008). Neural substrates of spontaneous musical performance: an fMRI study of Jazz improvisation. *PLoS One, 3*(2), e1679.

Liu, S., Chow, H. M., Xu, Y., Erkkinen, M. G., Swett, K. E., Eagle, M. W., ... & Braun, A. R. (2012). Neural correlates of lyrical improvisation: an fMRI study of freestyle rap. *Scientific Reports, 2*, 834.

Other

Seifert, G. (2009). *Langzeitpotenzierung in der CA1-Region – Verschiedene Modulatoren.* Inaugural dissertation. Freiburg: Medical Faculty of the Albert-Ludwigs-University.

Chapter 5
Research Question

The question to be derived from the gap in research presents itself in this way:

"Can it be achieved by integrating ACT-R, *Integrated Theory of Mind*, the four-level model of personalities and the model of hierarchical control based on the REST state, as well as the CREATES model of creativity, to find an explanation for the question of how something new and thus creativity can emerge from thinking?"

To answer this research question, it is also necessary to ask the following questions:

- What common elements connect ACT-R, Integrated Theory of Mind, the four-level model of personalities, and the hierarchical control model?
- How can the integration of ACT-R, Integrated Theory of Mind, the four-level personalities model, and the hierarchical control model be accomplished?

K. Testor, *Cognitive Theoretical Foundations of Creativity*,
https://doi.org/10.1007/978-3-030-61736-3_5

Chapter 6
Methodology

Contents

The methodology of answering the research-guiding question is a theoretical one, which creates a deductively derived modelling combination from the models mentioned, which is then examined in further consequence on the basis of various experiments of creativity research or adjacent areas. The author considers the disciplines framing the reference models as "micro worlds" as is seen in constructive realism, the obligations of which apply within this framework, and can only be imparted by "strangification" (cf. Greiner 2013, p. 5). In this sense, the parameters common to these areas in the structure should also be seen, which can be worked out in unison or at least similarly across the different approach worlds. The constructs that are thus to be classified as binding because they are necessary in different models (cf. Guttmann 2000a, p. 85), then integrate into the model used in this work.

Thus, it is important that the parameters necessary for the performance of creative processes are related to the fundamentally existing parameters of cognitive processes on both the anatomical and the abstract higher level. This distinction will specify a model of thought and neuronal interaction that should provide an explanatory approach to the experimental results to which it will be related during the investigation. At the same time, this answers the sub-questions of the research-guiding question.

The approach in the sense of a *strangification* will be carried out as less than an in-depth analysis by separating it from its own context and placing it in a foreign context (cf. Greiner 2013, pp. 8f., 10f.), as this would go beyond the scope of the work. However, the work nevertheless follows the direction that by referring to the

K. Testor, *Cognitive Theoretical Foundations of Creativity*,
https://doi.org/10.1007/978-3-030-61736-3_6

similarities of explicitly and implicitly used aspects, an approximation to a more general approach is to be achieved (cf. Greiner 2013, p. 9), because the different approaches adopted for the different "micro worlds" are similar, but not the same. Accordingly, an approximation to the more general binding nature can be attempted insofar as this will take place in the sense of an interpretation, that is, a change in the basic material, which thus becomes a "strangification" (cf. Wallner 2002, p. 233). Although the level of the object is not abandoned, nevertheless these findings are low in the structure of the administration of science. Thus, the results of this work are to be understood under the conceptuality of constructivism and, consequently, only in the *realities* of the sciences concerned (cf. Wallner 2002, pp. 134, 204, 213, 220f.).

6.1 A Model of Thinking

To understand thinking, it is necessary to work out the various areas from the models mentioned (ACT-R, Integrated Theory of Mind, four levels of personality, model of hierarchical control), which, as described in more detail in Chap. 4, are oriented towards specific questions and thus serve the focus of this work in partial aspects. In their integration, approaches to answering the research-guiding question can be expected. Nevertheless, we must first briefly outline another aspect that may be of great importance for the understanding of thinking and experiencing.

6.1.1 Constructs

> The "insight that we have learned to recognize all our categories of description of mental phenomena as more or less useful constructs, created by us to describe, predict, order, and classify where necessary for a particular purpose," according to Guttmann, is "the most remarkable advance in psychology" (Guttmann 2000a, p. 84).

Constructs that categorise the person-created *reality* are expressed not only in science, but also in a person's entire life. Guttmann was able to confirm this insight in an experiment he conducted. The test subjects were exposed to a longer presentation of exactly the same loudness and tone height. They were instructed to pay attention to slight differences in volume and to detect them by pressing a button. This experimental arrangement was tested by means of EEG potential evaluation and showed that the potentials of the lower experienced tones were reduced in amplitude whereas those of the louder experienced ones were amplified. Because this took place with the physically objectively same impulse stimulation, a discrepancy experience is expressed here, which is very useful for the understanding of the constructivist approach; this is confirmed scientifically (cf. Guttmann 2016, p. 1f.).

The author also sees the interconnection structure of the visual sensory cells as a further insight that makes this difference clear; these cells already generate a reduction of the physical sensory impression through mutual interference. This idea has previously been explained using lateral inhibition (Sect. 3.2.1) and is shown in Fig. 3.4 by the distinction between peripheral stimulus and central contrast. This already shifted reality favors a conception of the world based on constructs of abstraction and its interaction as well as an understanding of the world. This knowledge must therefore be part of the cognitive models and at the same time influences the understanding of the use of the term *representation*, which is frequently used and associated with memory content. The representation in the sense of this work and of the author is *not* a representation of an environment, but a construction that is reflected in patterns of activity and which is associated with certain external stimuli in space–time. It is therefore not a variant that can be derived from an input–output theory, but a view based on the theories of Maturana and Varela that an organism and its nervous system (re)act on these situations through externally triggered perturbations to maintain a specific inner balance in a variation of free drift. Thus, it is not the environment that produces the organism, but the organism itself and its reactions caused by perturbations (cf. Maturana and Varela 2015). To consider this is an important aspect, because otherwise the insight of the construction of the living world would be ignored and misjudged. This is explained in more detail in Sect. 11.1.

6.1.2 Common or Shared Terms

The terms discussed in this subchapter are part of the models mentioned in Sect. 3.3, and are examined here for their implicit or explicit aspects to confirm a broader meaning on the one hand and to be able to carry out a relativisation in certain areas on the other hand.

6.1.2.1 Chunks

As already described, a “chunk” is a representation of declarative knowledge, which contains many associations that have been created in large quantities by encoding the environment and are thus available. The additional references to the specification of the category are so-called isa-pointers (cf. Anderson 1996b, p. 355f.).

Through the relationship structure of the chunks, they are connected in a cascade-like manner (Anderson 1996b, pp. 120f.–127f.), because a term can serve both as a superorder and in an association on the same level or as an integral moment of the understanding of another term. Through this networking use, activation has a special significance, which is taken as a basis by other framework conditions (cf. Anderson 1996b, pp. 58, 62, 65).

In the Integrated Theory of Mind, chunks are also the substrate of environmental encodings in declarative memory. In contrast to the ACT-R, however, these are no longer called up individually, but compete with each other in the background, which they negotiate through topicality, context reference, and a cost–benefit calculation (cf. Anderson et al. 2004, pp. 1036, 1042).

In addition, chunks are ascribed purely explicit content. The implicit ones are reflected in production rules of the sub-symbolic processes (cf. Anderson et al. 2004, p. 1044).

The cascade-like networking of chunks can also be found in the structure of language, namely, both in the phrase structure (cf. Anderson 1996a, p. 346) and in the pause structure derived from it (cf. Anderson 1996a, p. 350f.).

Chunks as a unit of meaning of memory contents are also a moment of essence for Roth and Strüber. After processing and comparison with the known, the perceived is combined to form a unit of meaning: this includes the development of multiple internal levels of representation (cf. Roth and Strüber 2014, p. 217). Here, too, the combination of different contents into one unit is understood.

By the general structuring of the model of the four levels of personality, which are hierarchically connected, the derivations of the cascade-like connection of the contents can also be understood here, which however experience the restriction that some are generally accessible, whereas others however are not (cf. Roth and Strüber 2014, pp. 218f., 210). This information can also occur both in language and in images (cf. Roth and Strüber 2014, p. 304).

Rösler also presupposes chunks as a composition of the networked contents of specific environmental representation. On the one hand, the entire brain is involved in the activation of a word meaning. On the other hand, the distribution of the activation patterns determinable by voxel is similar in meaning. Rösler believes that there is a certain proximity depending on the filter function,[1] that is, that the perceptual basic features express the weightings in the meaning of language (cf. Rösler 2011, p. 189). Thus, the similarity with the understanding of the integration of environmental stimuli into memory content existing in the other models is strongly given. For this term, Rösler uses the expression *lexical access* and connects it with the concept coined by Anderson, on the one hand by describing the associative quality of the memory contents through their "fraying into other nets" and on the other hand includes contextual dependency and denotative and connotative meanings (cf. Rösler 2011, p. 371).

The significance of these chunks in the model for hierarchical control is also derived as an implicit prerequisite, as Rösler himself says that the model is likely to be extended like a cascade from higher levels to lower levels (cf. Rösler 2011, p. 342). It is therefore not only action- and process oriented but should also be a model for the general adjustment of competing activity patterns, which is sufficient to enforce one pattern against the other. However, if strictly implemented, contextuality as a prioritizing activating factor in one situation is jointly responsible for the enforcement of one chunk against another. Therefore, chunks are not only present,

[1]A filter function is the functioning differentiation referring to relevant contexts. Exemplarily, words such as mouse and cat are closer associated to mammals than the verbs run or go. However, when the frame is changed to an observation of movement, the tides would turn.

but at the same time prerequisite of this model, which otherwise could not produce anything specific without content. Rösler expresses this as follows (p. 343):

> "This regulation is unspecific in content insofar as the frontal areas merely register the relative activation of the content-specific representations, compare them with each other and finally reinforce or weaken them, without the entire content information being represented in the frontal areas."

In summary, it can be stated that chunks represent a complex representation of sometimes associative environmental stimuli, which express themselves in the brain through the formation of a specific activity pattern and that are connected to each other in different cascade-like structures through the context-dependent relationship between them.

6.1.2.2 Traces of Memory

In Anderson's work, the traces of memory are the pre-excitations of content activated by associations emanating from memory contents, which activate associative concepts (cf. Anderson 1996a, p. 178). Through these properties, phenomena such as priming, which explains a faster and more probable recall of a theme/word/etc. activated by preparatory stimuli, become understandable with the preparatory activation. Priming is the activation by the triggering stimulus, which should spread along the learned traces of memory (cf. Anderson 1996a, p. 182f.).

Traces of memory depend on exercise and are kept active longer the more they are used (cf. Anderson 1996a, p. 198).

In his two theories, ACT-R as well as in the *Integrated Theory of Mind*, Anderson does not address the traces of memory as directly as in his book on cognitive psychology, but he sees this concept as "corresponds to my ACT theory" (cf. Anderson 1996a, p. 178).

Traces of memory as an element of the associative networking of neuronal information are also an implicit construct on the model of the four levels of personality. Thus, it is described there that the degree of memory of the preconscious depends on the type of anchoring, that is, its degree of detail, as well as on corresponding connecting factors, but also on other parallel contexts that result in the degree of networking (cf. Roth and Strüber 2014, p. 227).

Furthermore, an indirect confirmation of the traces of memory is explained by the connection pattern presented as the basis of the processing, that is, that neurons interconnect locally and connect to other populations via a few central neurons (cf. Roth and Strüber 2014, p. 216). This pattern shows that the activation of nearby neurons reaches other nearby neurons more easily than distant ones. However, the fact that neurons are not solely responsible for the representation of information, but only through their networking representation, has already been explained in Chap. 3, The voxel approach, but is also evident from the network-like and function-specialized character of the brain (cf. Roth and Strüber 2014, p. 42). More recent ideas, such as the consideration that the perineural network, which is important for synaptic transmission, has an important role in the storage of information (cf. Sanders 2015, passim; Underwood 2015, passim), are negligible factors here, because the principle of network interaction as well as the resulting preparation of activities along traces of memory remain unaffected.

Traces of memory are also implied by Rösler in his model of hierarchical control. Thus, Rösler expresses these indirectly through the shift of linkage weights emanating from Hebb's learning rule (cf. Rösler 2011, p. 31f.) and their effect on the linkage of semantic memory contents. One can derive these semantic memory contents in the strength of association from their frequency of association corresponding to reality (cf. Rösler 2011, p. 188). In it one can recognise the memory traces, which express a relationship strengthened after frequent association.

The model of hierarchical control does not consider that competing states contain information, because these contents constitute the model by detachment from specificity. Nevertheless, it can be stated here that the concept of memory traces should also be coherent for Rösler. Furthermore, the presence of memory traces in this model implies the cause that the two conflicting states have come to one level, namely, by the higher activation in their area compared to other states from this area. This concept also expresses a better networking via memory track, which would manifest itself in a longer active posture as well as easier excitability.

In summary, it can be stated that memory traces occur in all models and thereby express an associative property related to an activated content, which enables a faster retrieval for a subsequent use as well as an active retention, depending on the frequency of use, over a longer period of time. In addition, it is easier to link memories to this content than to other areas.

6.1.2.3 Production Rules

For Anderson, production rules are the second essential element necessary for the emergence of complex cognition and express procedural knowledge (cf. Anderson 1996b, p. 355).

Production rules follow a certain goal. However, they can also be separated into sub-targets and thus imply a hierarchical structure. Through continuous development, they also represent the evolving understanding of problems (cf. Anderson 1996b, p. 357).

Production rules are generated by the fact that they specify the changes of chunks and are based on experienced examples (cf. Anderson 1996b, p. 359).

As do chunks, production rules compete with each other and are selected by weighting and balancing them according to the same principles (cf. Anderson 1996b, pp. 360–364).

In the Integrated Theory of Mind, the production rules, of which only one can be active at any one time and which are located in the basal ganglia, are responsible for the selection of the chunks available in several buffers (cf. Anderson et al. 2004, p. 1036). At the same time, they are jointly responsible for updating the buffers, which happens in a cycle lasting about 50 ms. Because only one production rule can be active at a time, this is one of two bottlenecks in this model (cf. Anderson et al. 2004, p. 1038).

Representation in the basal ganglia must be present or it must first be constructed from the environment. To circumvent this limitation in the degree of freedom, Anderson has more closely defined the emergence of production rules, creating a more general approach: the production compilation. Here, pairs of successful production rules are brought together in such a way that they can then produce the effect of both (cf. Anderson et al. 2004, p. 1045).[2] This process is also consistent with expertise research, which sees the convergence of chunks into more meaningful units confirmed by an extension of their application (cf. Sect. 3.2.2.6), because production rules can be generated from the understanding of chunks and their change.

Roth and Strüber write in the formation of consciousness about preconscious processes, such as intuition, which take place just below the threshold of actual consciousness (cf. Roth and Strüber 2014, p. 230). Furthermore, through the differentiation of sensory stimuli in "unknown–known" and "unimportant–important" the intrusion of information into consciousness is already limited and regulated by processes (cf. Roth and Strüber 2014, p. 212). It can already be deduced from these two aspects that production rules exist.

The formation of "answer routines" expresses that the processes present in the basal ganglia are automated and are elements created from experience (cf. Roth and Strüber 2014, p. 212). This conclusion seems to give the impression that they could run in parallel. This is a topic that is currently still being discussed, seen both in a permanent activation of change between the tasks and in a quite possible parallel processing (cf. Fischer and Plessow 2015, abstract, discussion).

As Rösler's model of hierarchical control was based on the negotiation (in sense of bargaining) of optional decisions, that is, the selection of possible options, just as control itself is also a process, the production rules are also to be found here (cf. Rösler 2011, p. 337).

It is interesting to note that this control functions similarly as a special type of lateral inhibition at different hierarchical levels (cf. Rösler 2011, p. 342), as well as the increasing complexity of this cascade formation, which makes it possible to summarize sub-rules (cf. Rösler 2011, p. 324f.).

Because the basis of this model was the function under conflicts of action and thus a model for decision making, which is concerned with the functioning of the basal ganglia and the learning of routines in these (cf. Rösler 2011, pp. 265f., 268–276), the moments of the production rules are also given here.

[2]To stick to the ACT-R language, which is still the foundation for this model, a mathematical operation like an addition could be used exemplarily to explain this in detail. If you imagine the addition of numbers as the basis, the combination of additions could lead to a new production rule: the multiplication. If we now would take the automated movements approach in Anderson's Integrated Theory of Mind, we would come up with the combination of a leg extension together with bending the body to make a roll or a somersault. You could also combine leg extension and an arm movement to create a new production rule for gear shifting.

In summary, it can also be stated here that the production rules comprise a representation filtered from the environment and associated with it, which encompasses procedural knowledge and, at the same time, are able to represent more complex mechanisms in a simplified way by means of integrations, resulting from to the relationships existing among themselves. This integration also results in hierarchical cascade-like relationships, which are different in context. Analogous to the chunks, it can also be assumed here that these production rules go hand in hand with specific activity patterns.

6.1.2.4 Hierarchy Levels

The different levels shown by the hierarchical control model are the next aspect to consider. In this approach, which should be present in cascade form over the more complex abstract areas of the brain and the sensory processing primary and secondary areas (cf. Rösler 2011, p. 342), a decision process given by the circuit structure takes place. This level does not require any specific references, but by a process similar to lateral inhibition in a contrast and thus in a polarisation leads to two switching states for decision making: a blocking zero position and a letting through one position.

The already mentioned hierarchy is not expressed in a hierarchy of higher values and thus dominating all other areas, but in a function-related one. Thus, this hierarchy is already in the visual cortex a continuation of the stimuli structuring on lower levels, which lead to shaped contents through increasing networking (cf. Rösler 2011, p. 70). However, in this area no decision in the sense of behaviour has yet been taken, but one in relation to the structure is to be recognised. These autonomously distinct areas lead in the interaction on higher levels to form interpretations, which innervate specialized areas for different object categories, as, for example, the different areas for the recognition of houses or faces (cf. Rösler 2011, pp. 70f., 241f.). Thus, even on these lower levels, a decision takes place regarding the interpretation of visual stimuli, which place an activation of one area in front of the other and can thus activate the following cascades in preparation, as would be expected in the context of memory traces.

An important feature of these hierarchies is the lower level consisting of two neurons or neuron networks, which has an exciting and mutually inhibiting effect on a level above. This action creates a circuit in which the strongest activated setting on this level can assert itself as the dominant one and subsequently be activating for higher levels, leading to further adjustments or an implementation (cf. Rösler 2011, p. 337f.).

The restriction of these hierarchies is thus limited to those areas that show few declining connections. This limitation was shown, for example, by connectivity studies in the area of the dorsolateral prefrontal cortex, which is hardly connected to and within the amygdala, which explains why emotions are difficult to control or change by ratio (cf. Roth and Strüber 2014, p. 348).

But aspects such as attention show that this hierarchy is not only to be seen with one above and one below, but basically simply as a decision-making process that helps interacting and competing networks in mutual interconnection to dominance and thus enforcement, but is also likely to be influenced by other loops or processes.

The four levels of personality model can be seen both through the distinction between conscious and unconscious levels and through the specialisations emanating from these levels with a similar approach to the understanding of the hierarchy of neuronal processes. For example, Roth and Strüber write that despite the hierarchy, the connection of primary and secondary areas in the visual cortex with higher associative areas is necessary to allow stimuli to penetrate consciousness (cf. Rösler 2011, p. 225). This is where the bargaining processes are happening, which they also call the decision between *known–unknown* and *important–unimportant* for stepping into consciousness (cf. Rösler 2011, p. 212).

A further indicator for the existence of these hierarchies is the connection pattern of the cortex, which is divided into local and central connections by them (cf. Rösler 2011, p. 216f.), and thus might imply a stepwise specialized interaction of neuronal associations, as shown by the approaches for explaining the mode of action of psychotherapy. These approaches to the mode of action are not seen directly top-down, but as described, can be reached via detours into the area of the amygdala (cf. Rösler 2011, p. 367).

For the ACT-R as well as for the Integrated Theory of Mind, it has already been stated that the chunks as well as the production rules are cascadingly connected, whereby this connection can shift by the context reference. As Anderson applied this model in simulation, he solved the negotiation of these competing knowledge contents in specialized calculations (cf. Anderson 1996b, pp. 356–366; Anderson et al. 2004, pp. 1042, 1044f.).

Furthermore, he is of the opinion that "this strong abstract, hierarchical structure on behavior," which makes human cognition special (cf. Anderson 1996b, p. 357). By this he means that production rules follow a certain goal. However, they can also be broken down into sub-goals (cf. Anderson 1996b, p. 357) and imply a hierarchical structure. Here, the hierarchy expresses through the necessity to assert oneself that these are competing processes, which are then negotiated.

In addition, the cycle of about 50 ms that he cited, which concerns an adaptation of the content focused in the working memory, in terms of both declarative and procedural knowledge, through a reevaluation is coherent (cf. Anderson et al. 2004, p. 1044) to the extent that the possibility of maintaining presence and changing competing states mentioned by Rösler is depicted (cf. Rösler 2011, p. 342).

In summary, it can also be said here that the hierarchy levels are present in all models as a negotiation mechanism through mutual interconnection and are thus capable of making decisions about situational processes through the generation of switching states. These decisions can, however, lead to a change in the decision at any time by shifting factors of the lower levels. These lower levels are to be understood as the context- and content-dependent cascade formation, which can vary. Because of this difference in relation, other processes, such as attention, also intervene in these negotiation mechanisms and can cause a weighting shift in favour of one side.

6.1.2.5 Consciousness

Consciousness is a "modular, spatially and functionally separated, albeit under normal circumstances closely interacting organization of the different contents of consciousness" (Roth and Strüber 2014, p. 210).

Consciousness is relevant to this work through the four-level model of personality, which distinguishes between the conscious and the unconscious. The dividing line is shown by the fact that a threshold must be crossed for becoming active in consciousness. And it is true that the slower something becomes conscious, the more "a stimulus situation is more complex, unexpected or unclear in its meaning" (cf. Roth and Strüber 2014, p. 202). But even if some signals are too weak to be brought into consciousness, they can still influence behaviour and trigger emotional or behavioural changes via the thalamus in the activation of the amygdala via subcortical processes (cf. Roth and Strüber 2014, p. 201).

Consciousness is necessary for the processing of new contents, the processing of events of greater complexity, the merging of different contents and their anchoring in declarative memory, the meaningful processing of experiences, and the planning of complex actions in new situations as well as the rapid prediction and performance of complex social interactions including language (cf. Roth and Strüber 2014, p. 211). Roth and Strüber define consciousness from a psychological point of view as "format for processing new, important and meaningful information [...], whether it is [about] perceptive, cognitive or emotional events.," which is a prerequisite for storing in long-term memory and cannot be achieved by simple intuitively working centers (Roth and Strüber 2014, p. 215).

It is strongly related to the working memory and the "global workspace," because from a psychological point of view this is the place for action planning (Roth and Strüber 2014, p. 214), and there occurs a "tremendous complexity reduction" (Roth and Strüber 2014, p. 215), which would be expressed in the focus of consciousness reducing only to one aspect.

Consciousness is therefore needed to transmit external stimuli into the associative cortex areas and to store them subsequently (cf. Roth and Strüber 2014, p. 201). Furthermore, increased attention is important for the degree of memory of meaning, as is emotional excitement. Everything conscious sinks, if it is not kept in the actual consciousness, into the preconscious (cf. Roth and Strüber 2014, p. 227).

The subconscious "contents of long-term memory which are currently not conscious, but which are based on certain perceptions, experiences, messages and the like, that can become conscious 'by itself' or through active remembrance" (Roth and Strüber 2014, p. 227).

For Rösler, consciousness is not an explicit topic, because it is not only about the approach of self-organizing dynamic systems in decision behaviour (cf. Rösler 2011, p. 265) but also in the abstraction of statistical methods on which processing is based as "neuronally generated" by accumulated evidence "likelihood-ratios" (Rösler 2011, p. 247ff.) and works with automatic processes.[3]

Nevertheless, approaches to the meaning of consciousness can also be recognized in his work, although this must be sought in the treatment of attention. It can be seen that he sees both the effect of attention as a supporting element of the target action and its connection with the contrast effect caused by attention (cf. Rösler 2011, pp. 94, 97). He also expresses this as follows:

> "Selection as an 'orientation of attention' means that, on the one hand, certain stimuli of our environment are placed in the centre of consciousness, while at the same time many other stimuli are completely ignored or at least weakened to such an extent that they do not disturb us in observing the contents that are the focus of attention" (Rösler 2011, p. 79f.).

This quotation shows that for Rösler, consciousness can be expressed in attention.

With Anderson, the concept of consciousness is also not explicitly present. But because he assumes that main characteristics as parts of an object are already available before recognition, but only create an unambiguous identification with the focus of attention (cf. Anderson 1996b, p. 359), attention is also found here as an implicit element of expressed consciousness.

Furthermore, the procedural and declarative processing of Anderson, brought together in working memory, is covered in both theories, which imply the existence of consciousness in this processing. This concept should also be seen in the light of the fact that Baddeley refers to the superordinate control and attention function (cf. Gramann 2002, p. 6) designated as the central executive, which thus carries awareness and purposeful processing within itself through attention.

In summary, it can be said that consciousness in the neurosciences is associated with the attention that can be better found and measured in biological correlates, but is generally seen as a prerequisite for neuronal processes, if these are to be sustainable and retrievable. In addition, consciousness, when expressed in attention, can influence a shift in activity intensity. The consciousness is probably to be found in parts of the working memory, which is influenced and controlled by it.[4]

6.1.2.6 Context

The influence of context on the storage of information and thus on the chunks was already derived in Sect. 3.2.2.3, The influence of contextuality as the current state of science. Because, from the author's point of view, this aspect could make a

---

[3] Probably Rösler's position, that the recording of biological correlates only measures the effects of single cells, and therefore a detection of a consciousness as well as its origin cannot be captured, is one reason for not including this topic specifically (cf. Rösler 2011, epilogue).

[4] It has to be explicitly stated that the author intentionally uses the recordable effects of consciousness and its connection with aspects of mind in relation with the mentioned models and theories only, but excludes, as already insinuated in Sect. 2.3, Neurosciences, the more general question about the body–soul dualism as well as questions directly linked to it.

significant contribution to the creation of new things, the models that give reference are now considered from this point of view.

In the ACT-R as well as the *Integrated Theory of Mind*, contextuality has an important role. An important aspect in the formation of chunks, for example, is that the signals coming from the environment acquire an internal associative character through the reference setting (cf. Anderson 1996b, p. 359). Because these associative contents are not defined in more detail but define themselves via the isa-pointer as a more comprehensive description and delimitation (cf. Anderson 1996b, p. 356; Anderson et al. 2004, p. 1042), which represent statistical frequencies (cf. Anderson 1996b, p. 361f.; Anderson et al. 2004, pp. 1042–1045), this implies the somewhat more complicated application of a context relatedness.

The context leads to physically identical stimuli being interpreted differently in different environments and thus causing different decodings and associations in the respective framework conditions (cf. Anderson 1996a, p. 58f.). Only a product provided with context becomes the basis for higher cognitive processing (cf. Anderson 1996a, p. 71).

In addition to the chunks, the production rules are also designed with the environment in mind, which is done by encoding changes in the environment (cf. Anderson 1996b, p. 355). Thus, chunks and production rules are both dependent on the surrounding contexts, in terms of both the environment and the interactions taking place there. Even the internal new production of production rules does not dissolve this reference, as it is based on the effects of the previous ones (cf. Anderson et al. 2004, p. 1045). Only the rule of production changing in the following experiences through other context relations could detach itself from it, but what still happens through a context relation could also include the predecessor rules in the change.

The context is an important point for retrieval, because it activates the related knowledge, that is, the knowledge corresponding to the context, and thus enables appropriate action (cf. Anderson 1996b, p. 360).

For the four levels of personality, context is an important aspect in the coding of addresses for the contents of long-term memory stored in the hippocampus (cf. Roth and Strüber 2014, p. 82) and thus has a referential role; therefore, context learning is a secondary amplifier (cf. Roth and Strüber 2014, p. 319).

An important part of context learning, apart from the number of connecting factors, which can be expressed both in the richness of detail and in the number of different frames, is emotional excitement (cf. Roth and Strüber 2014, p. 227). Thus, the ventral medial prefrontal cortex is associated with the storage of the combination of physical excitation and the aspects of the situation which, when a comparable situation occurs, triggers a reactivation of the respective excitation pattern (cf. Roth and Strüber 2014, p. 287). Thus, the emotional frame of a situation represents a special context factor.

The influence of a specific environmental situation association can lead to a shift because of its experience under different framework conditions, as is the case, for example, with confrontation therapy in behavioural therapy. The influence of the context is therefore subject to the same adaptation mechanisms as learning itself and is likely to be expressed in both the LTP and the LTD at the neuronal level, with limitations in the area of fear learning (cf. Roth and Strüber 2014, p. 340f.).

Contextuality also has an important function in the hierarchical control model, leading to a convergence in the dorsolateral and ventrolateral prefrontal cortex of the context information stored from the sensorimotor cortex and the posterior, which forms the basis for the following activities at higher levels (cf. Rösler 2011, p. 322). The context thus becomes a basis for processing at higher levels and has an effect, for example, on the negotiation of action goals by influencing them by innervating the posterior association cortex (cf. Rösler 2011, p. 338).

In summary, one can say that context has a role in all models. It is inserted as an associative information important for memory content as well as for negotiation processes and production rules and essentially contains all activity patterns surrounding the information, including emotional activity. This insight underlines the understanding of the brain neuronal activity pattern as a vector, which can be expressed as a voxel of different resolution. At the same time, context is important in framing in both storage and retrieval, influencing the activation of different appropriate patterns and negotiation loops.

6.1.3 Summary

From the elaboration of the common terms, it can be seen that the different contents are characterised above all by their interconnectedness and thus their association. Both the memory contents and the processing structures refer strongly to each other and express a relationship structure. As described in Sect. 6.1.1, Constructs, these are not general representations but already abstract constructions.

The contents of memory, both declarative and procedural, are created by sensory inputs. Statistical patterns of negotiation and distribution are essential in this process, which subsequently lead both to the association of these patterns and through them to a reduction. These elements are adaptive and can adapt, although there seems to be a limitation of this plasticity in certain areas.

Through the contextuality as well as the frequency of use, these networking patterns become more stable and form structured expansion patterns/memory traces, depending on the framework conditions.

General patterns in conflict with each other are brought into the on or off position by contrast, whereby the “off” does not mean that they have been rejected, but only that they have receded into the background.

Consciousness, which can be recognised among other aspects as attention, changes thoughts, neuronal activation patterns, or conflicts in such a way that shifts are possible which can affect both internal and external processes.

The interconnection structure of the brain is partially changeable, both in the cross-linking density of synaptic connections and through adult neurogenesis, but is based on a basically genetically predisposed circuit.[5] The extent to which this epigenetic change, and thus the transmission of information, encompasses, is a question that also takes on new significance, for example, in the investigation of transgenerational trauma.

Both internal and external sensors and receptors are understood as sensory stimuli, as these are always perceived as receptors by the brain and thus have an influence on neuronal contextualisation. This idea was shown, for example, in the severing of the vagus nerve in mice when they no longer reacted to bacteria in the intestine with stress-reducing effects[6] (cf. Cryan et al. 2011, abstract). Thus, this sensory input, which is innervating from this area and especially important for this aspect, was interrupted, which indicates that internal areas can also be regarded as sensory. The brain thus makes a stringent distinction between sensory inputs and internal circuitry.

6.1.3.1 The Brain as an Organ of Association Representing Constructions

The brain is an organ that brings the sensory information reception from the environment as abstractly reduced contextualised constructs of reality with associative parts into a representation, relies on a partly genetically predisposed and partly adapted neuronal circuit pattern with specialised areas and loops, and whose processes organize themselves basically automatically by self-organisation. Thus, this information is referenced, reflected, and (re-)constructed.

In short, the brain is a receiving associative organ that constructively represents the environment.

The attribution of the soul or life energy that initiates and triggers these processes is unaffected by this and remains unnoticed because it is even less tangible than consciousness.

[5]You can derive this insight from cataract studies, which show that an eye functionally repaired later in life leads to vision competence in people otherwise blind since birth (cf. Chatterjee 2015, passim). In other people blind since birth, the visual cortex limitedly interconnected with the tactile or language processing areas (cf. Rösler 2011, p. 154). The possibility to gain sight again implies that the formerly existing structures remained and did not get lost, and it also means that the projections and areas in use from other specialized fields reduce to give the necessity of connectional potential free for visual processes. These points add up to the conclusion, that the (genetically) predisposed specialisations in the neuronal network rank higher in priority to others, even if these others already established a functional working extension.

[6]A bacterially (*Lactobacillus rhamnosus*) induced receptor expression of GABA in the central nervous system calmed fear reactions as well as depression sensitivity in mice, influencing their behaviour, as Cryan was able to show in predecessor studies.

6.1.3.2 Thought Definition

Because the brain is the biological functional basis of thinking, which is expressed in the circuit structure as the basis for neuronal processes, the thought should be able to be worked out from this interaction and express itself as some kind of biological correlate.

As it has already been recognised that through contextualisation all areas of the sensory inputs lead from the lower levels into converging areas, express themselves there in different activity patterns depending on the attribution of the content as well as on the attention and thus address different filters, these aspects are to be named as the external part of thought.

The counterpart is the internal thought component, which connects all previous chunks and production rules through internal networking and represents activation caused by external stimuli, taking into account the regulating intervention of consciousness.

The connections of these two parts of thought can be represented vectorially as snapshots by means of voxels: this makes them unique. At the same time, this activation of the two components enables neuronal plasticity.

As consciousness is an important factor for the thought, no matter in which state of consciousness intensity one is at the moment, it has to be included in the definition of a thought. The attention supports a relative increase of the focused part compared to the others.

As a result, a thought is a special vector constellation of the neuronal activity pattern generated by an external and an internal part, which is perceived in the consciousness. Through the activation of consciousness, the system is sustainably influenced.

6.2 Creativity in Relation to This Model

Although this now roughly outlined model of thinking describes how thoughts can be (re)produced in it from previous knowledge, the question for the connection with creativity is now how the state of free association could be carried out in this system, as Andreasen described it in her REST state. It must also be clarified how the connection will be established with the CREATES model.

6.2.1 REST State

For Andreasen, the REST state is a state derived from episodic memory, which takes place in its reference to current spatial and temporal circumstances and thus offers the possibility of introspection, which is not the case in the generalised abstract of semantic memory (cf. Andreasen 2005, p. 71).

The associative areas of the inferior frontal lobe as well as parietal associative regions and the associative cortex are named by Andreasen as active areas for this condition (cf. Andreasen 2005, p. 73). An essential element is that the omission of the focusing areas enables free association and thus a freer integration of knowledge content. New things can be formed through these new connections.

The REST state therefore means above all a reduction of concentrated attention.

6.2.2 *CREATES Model*

The CREATES model contains the essential statement that in the process of creative work different phases with special brain activities can be better solved and that there are also different approaches for this process. Thus, a connection can be established with the developed aspects of the model of thinking.

An important distinction here is that the process is subdivided into both new creating elements and focusing and integrating elements. This differentiation now helps to obtain approaches for the investigation of creativity, taking neuronal principles into account.

The main factors according to Carson are cognitive disinhibition, activation of the right hemisphere; mood and mental orientation in active or passive are listed as key differentiating factors (cf. Carson 2010, p. 267).

Furthermore, it can be generally inferred from their processing that the associative areas are of particular importance. Integrating these will thus become one of the points to be examined in more detail in the study (cf. Carson 2010, pp. 114, 165, 246).

Additionally, it becomes apparent that the semantic areas are important when it comes to the free flow of thoughts in the connect state as well as the reference to experience as a basis and memories with these flowing thoughts, as is the case in the envision state (cf. Carson 2010, pp. 105, 113).

A further aspect is the change of the evaluating areas in phases of creation, because these can otherwise have a regulating and thus blocking effect (cf. Carson 2010, pp. 123f., 155–160).

This aspect makes it possible to establish a connection both with the importance of adaptation in attention, as was the case in the REST state, and with the importance of promoting the associative parts of the cortex, taking into account all the information received and processed.

6.3 Recapitulation

As now explained, the model of thinking represents a rough structure that, combined with the approaches from creativity research, can show how something new can emerge from this model.

As starting points for the creation of something new, the chunks, the rules of production, the hierarchical elements, the conscious and unconscious parts, and the anatomical interconnecting structures as well as the traces of memory are possible.

These approaches are now further investigated within the framework of this work. On the basis of this model of thinking, there are several possible ways of creating something new. On the one hand, attention could be switched off or reduced as a limiting element, as would be the case in trance, hypnosis, or half-sleep, for example. Furthermore, drawing attention to areas that are otherwise not so active would be an effect that better links these areas, which could result in alternative connections. Another possibility would be to connect unused areas by active provocation to create such a connection and stabilize the evoked patterns that begin. Similarly, coincidental networking could be conceivable, which occurs within the areas because they happen spontaneously.

The definition of thought presupposes that this new network must be so strong that it manages to penetrate into consciousness, that is, it crosses the threshold from preconsciousness to consciousness.

The differentiations captured by these observations are now related to experiments carried out in research, leading to confirmation or rejection of the approaches, thus better defining the framework for the emergence of something new through the neural network.

The final aim is to summarise these findings in a categorisation. Thus, this work can be described as theoretical, which creates through deductive analysis a thought model with the ability to create something new and examines this in the course of an inductive control on the basis of different experimental arrangements as reference instrument in the reliability of the aspects worked out from the starting points of the model.

References

Books and Journals

Anderson, J. R. (1996a). *Kognitive Psychologie. Eine Einführung.* (2nd ed.). Heidelberg: Spectrum Academic Publishing House Ltd.

Anderson, J. R. (1996b). ACT. A simple theory of complex cognition. *American Psychologist, 51*(4), 355–365.

Anderson, J. R., Bothell, D., Byrne, M. D., Douglass, S., Lebiere, C., & Qin, Y. (2004). An integrated theory of the mind. *Psychological Review, 111*(4), 1036–1060.

Andreasen, N. C. (2005). *The creating brain. The neuroscience of genius.* New York/Washington, DC: Dana Press.

Carson, S. (2010). *Your creative brain.* San Francisco: Jossey-Bass, A Wiley Imprint.

Chatterjee, R. (2015). Out of the darkness. Defying expectations, cataract surgery in Indian children in Indian children is endowing them with vision-and shedding light on how the brain learns how to see. *Science, 350*(6259), 372–375.

Cryan, J. F., Bravo, J. A., Forsythe, P., Chew, M. V., Escaravage, E., Savignac, H. M., ... & Dinan, T. G. (2011). Ingestion of Lactobacillus strain regulates emotional behavior and central GABA receptor expression in a mouse via the vagus nerve. *Proceedings of the National Academy of Sciences, 108*(38), 16050–16055.

Fischer, R., & Plessow, F. (2015). Efficient multitasking: parallel vs. serial processing of multiple tasks. *Frontiers in Psychology*. https://doi.org/10.3389/fpsyg.2015.01366.

Guttmann, G. (2000a). Die Messbarkeit des Erlebens und der Irrtum mit dem Ich. In F. Stadler (Ed.), *Elemente moderner Wissenschaftstheorie*. (pp. 73–91). Vienna/New York: Springer.

Guttmann, G. (2016). Erfinden statt entdecken. In N. Holzenthal (Ed.), *Constructing reality/ Realität konstruierend* (pp. 25–32). Frankfurt am Main: Peter Lang Publishers.

Maturana, H. R., & Varela, F. J. (2015). *Der Baum der Erkenntnis* (6th ed.). Frankfurt am Main: Fischer Paperback Publisher.

Rösler, F. (2011). *Psychophysiologie der Kognition. Eine Einführung in die Kognitive Neurowissenschaft*. Heidelberg: Spectrum Academic Publishing House.

Roth, G., & Strüber, N. (2014). *Wie das Gehirn die Seele macht* (5th ed.). Stuttgart: Klett-Cotta Publisher.

Sanders, L. (2015). Nets full of holes catch long-term memories. *Science News, 188*(10), 8.

Wallner, F. (2002). *Die Verwandlung der Wissenschaft*. Jandl, M.J. (Hg.). Hamburg: Dr. Kovač.

Online Resources

Underwood, E. (2015). *Mysterious 'holes' in neuron net may help store long-term memories*. Washington, DC: American Association for the Advancement of Science. http://www.sciencemag.org/news/2015/10/mysterious-holes-neuron-net-may-help-store-long-term--memories. Accessed 27 Dec 2015.

Other

Gramann, K. (2002). *Working memory and electrocortical activity – an experimental study on the differentiation of subsystems of working memory*. Dissertation. Aachen: Faculty of Philosophy, Rheinländisch-Westfälische Technische Hochschule Aachen.

Greiner, K. (2013). *Constructivist realistic psychotherapy science*. Vienna/Paris: Teaching Script of the SFU Sigmund-Freud-Privatuniversität Wien/Paris.

Chapter 7
Investigation

Contents

As already mentioned in the previous chapter, there are now several connecting elements on the basis of the model that make it possible to approach the question of creation of new elements. It should be remembered that this means the creation of something new, from the point of view of both ordinary and extraordinary creativity. What is important here is the distinction between something new, which is useful, and something that does not bring any benefit, which is solid from creativity research.

From this understanding the aspect of the evaluation (in sense of assessment of value) penetrates, which is provided by the comparison group or reference society and which can change also over time, as this was the case, for example, with a van Gogh. Thus, the concept of creativity comprises "the intersection of the terms novelty and value. The distinction between new and old is, however, not rigid, but itself a situational and social process of permanent redefinition." (Brodbeck 2006, p. 5).

To exclude the aspect of assessment of value, the parameters necessary for the emergence of something new are not seen in the sense of the aforementioned understanding, but under the consideration of the general change of neuronal processes and interconnection structures that are assumed to be the basis for the creation of something new. At the same time, the important prerequisite for creativity, that something new emerges, is brought into focus to be able to work out the fundamental aspects of this process. The omission of the evaluation in general is at the same time also the reduction of arbitrariness, because evaluation is always arbitrary, depending on individual attributions. This point also includes the individual attributions of groups/societies, etc. beyond a single individual, provided that these can be named as an abstract whole. Last but not least, the process can be considered at the

K. Testor, *Cognitive Theoretical Foundations of Creativity*,
https://doi.org/10.1007/978-3-030-61736-3_7

grassroots level and a rating-oriented posture/objective oriented could be subjected as a subsequent step.

Following this train of thought, the individual aspects of the model of thinking on the possibility of creating of new ideas or their influence on such a process are now considered.

7.1 Segmentation and Consideration of the Model of Thinking

The different approaches mentioned in the method for a consideration of the creation of something new are covered in this chapter to create the basis for the connection to experiments.

7.1.1 Perception

Humans record their environment on different channels, which move through the organs of perception in a certain spectrum. Thus, the visual apparatus enables the detection of the so-called visible light spectrum, which is defined by the wavelengths between 400 and 680 nm. In this area, the eye has specialised cone cells that enable the processing of this light and which are specialised in specific areas of this spectrum (cf. Silbernagl and Despopoulos 2003, p. 356f.) This receptivity is at the same time a limitation that restricts humans in their ability to construct the world. Only through the use of aids such as UV or IR wave processing devices were these adjacent areas made visible and thus usable. Thus, through aids, an extension of the possibilities of understanding and thus the creation of something new was stimulated.

In addition to the visual channel, the sensory channels also include the auditory, olfactory, gustatory, and sensitive[1] channels. Each of these channels has a limit to its ability to focus on a specific area of the perceptible. So, the environment affects the body through these stimuli and expresses itself in it. By shifting the attention, these very numerous incoming stimuli, about 10^9 bit/s, reduced consciously perceived, about 10^1–10^2 bit/s (cf. Silbernagl and Despopoulos 2003, p. 312).

These perceptions are then passed on through the recording and transmitted to higher abstracting areas of the brain. This perception expresses not only the content focused by the consciousness, but also all other stimuli that are transmitted through these channels. They represent, so to speak, the context surrounding the focused element.

[1] Sensitive includes the tactile as well as the haptic perception: as tactile sense, these known parts usually comprise external as well as internal receptors and consequently create a multispectral system, which perceive temperature, pressure, position, etc.

These stimuli coming from outside, however do not only include real outside, but also inner outside. This refers, for example, to the intestine, which, as an area within the body, nevertheless represents its delimitation from the outside world. This is therefore of particular importance in the immune defence (cf. Silbernagl and Despopoulos 2003, p. 232) as well as in the sensory processing of information, which, if disturbed or deprived, can lead to depression-promoting states (cf. Wolf et al. 2016, in brief and passim; Prinz et al. 2015, passim; Cryan et al. 2011, passim).

As shown in the study by Wolf et al. (2016), intensive administration of antibiotics reduces adult neurogenesis in the hippocampus. Wolf and his colleagues were able to stop and reverse this process with a probiotic treatment. This finding shows that the formation of new neurons in the brain can be influenced by external stimuli. From this it can be deduced that in deprivation—the antibiotic state differs from the probiotic state among other things in a reduction of the plurality in the intestinal flora as well as in the quantity and thus probably represents a reduced number of stimuli in this area—sensory stimuli influence plasticity.

For the creation of something new, the approaches can be found here that this can be created both by changing the stimuli, qualitatively or quantitatively, or by focusing on the stimuli or the reference system of the stimuli to each other. This should be reflected in the next level of processing, creation of chunks.

7.1.2 Chunks

Chunks are the units of declarative knowledge: these include the referenced knowledge and express themselves in the association of other chunks as well as in their own immanent elements.

As already described in the approach to perception in the previous chapter, these are made available through various channels. Chunks are the reference of these environmental stimuli to each other as well as to the context of what happens in the corresponding convergence zones of the brain.

An exemplary element from the findings of cataract studies is the connection with the birth-blind people and language. Speech itself is absorbed and processed by the ear, allowing a hierarchical processing of the sounds as well as the words, sentences, or even texts to be shown. These different levels were used by cognitive psychology to test different theses and theories and to apply them in experiments (cf. Anderson 1996a, Chaps. 11 and 12). So, it turns out that words are derived from the sounds and are combined higher and higher into larger and larger sense formations.

The reference to seeing is expressed by the fact that the attribution of a name to one and the same object is different in different languages. For example, in German the word "Papa" or "Papi" is used in English as "dad" or "daddy" for the person named as the father.[2] The attribution of this person takes place through the sensory

[2] Similarly, the same would apply for mama and mum, but was not used because of the less obvious difference referring to (phonetic) spelling.

channels with the references to it. Thus, a baby learns this term through the coming together of auditory and visual stimuli from which at some point the connection 'person is equal to daddy' is created: this is shown by the fact that there is a phase of mixing male persons as dads as well as names and man/dad, etc. (cf. Schenk-Danzinger 2006, p. 157). The reference of the stimuli heard as well as seen and their connection is thus a learned, not to say conditioned, aspect. Among other things, the dopaminergic system is important as it presumably influences the formability of the existing information when something new occurs (cf. Roth and Strüber 2014, p. 100).

However, the connection with the visual system is recursive, as demonstrated by birth-blind people, who experience an expansion of the speech areas when the visual cortex is not used through optical impressions. Thus, these areas are not lost, but are used by other areas. This finding also means that these areas must innervate to enable activation at all.

Because the cataract studies showed that these birth-blind people can use their visual areas again, it becomes clear that the optical impressions can be transmitted and that the basic interconnection structure has remained unaffected, even if it has only been used in a reduced but different form. This feedback to the visual areas makes sense insofar that what has been heard has to be related to what has been seen. This can also be applied analogously to the other sensory impressions, although these are less directly integrated, but rather via the brainstem and the limbic system.

The important derivation is that these areas present on the lower level are closely related to each other, which leads to a chunk through the compression of the information.

The chunks are a sum of activity patterns that are also related to each other. This relationship structure is expressed in terms of terminology, categorisation, but also in other framework conditions. As chunks as units of declarative knowledge are already an abstraction, they constitute the sum of the inherent information. A chunk of a word comprises both the written letters and the sounds associated with them. In addition, the significance gained from and attributed to environmental interaction is also stored. Chunk is thus the carrier of information on one level.

This thought therefore leads to the question of what is the sum of two chunks. Two carriers of information on one level, when merged, become one carrier of second information on two levels, one quasi-superchunk, the same as a lowest level chunk does with the sensory information: it brings them together into a construct of information. Thus, "chunk" is an abstract term detached from levels, which is characterised by the fact that it can contain more or less information as well as content depending on the selected focus. The difference is shown in the level of abstraction. Chunks thus represent an integration of sensory information in different depths and represent it in specialised areas.

This description of the specialised areas is reflected in Guttmann's spot activation, which describes a narrowly defined focused phase of thoughts that he was able to show in hypnosis experiments (cf. Guttmann 2013, p. 12). One thought, in the way already defined in this work, is likely to be particularly distinguished by the fact that it is expressed in a special form of activity, which may have been caused by attention or inner focus, in the illustration of imaging procedures.

This idea leads to a changeable cascade formation of the chunks among themselves. This cascade is expressed in the respective different reference system which is currently being applied. Thus, in a cascade relating to fruit, both the term "apple" and the term "pear" would have to be seen at the same level, namely as a subheading of the term "pome fruit." Alphabetically arranged, apple would come before pear. In terms of favourite foods, pears could rank higher than apples. The same can be said of the examples of "bird" and "feather," which, when animals are mentioned, would name the bird and only then the subcategory "feather" as its dress, but, in the bedroom reference system, would name the feather underpadding and list the bird as one of its suppliers.

From this it can be deduced for the creation of something new that it is to be expected that something new can be created by changing the level reference of chunks. Furthermore, it can be assumed that the different framework conditions activate different specific chunks, as confirmed by priming. A change in this activation of the specific chunks should therefore be possible by shifting the framework conditions as well as an alternative activation of another chunk of a framework condition created by an intervention.

Chunks are therefore variable and at the same time express associations in their activation. These associations are the connection and the relation of the chunks to each other. For example, the associations Anderson calls the isa-pointer are an expression of what was experienced with a term. This relationship is expressed in the traces of memory.

7.1.3 Traces of Memory

"Traces of memory" is a general term for the relationship between memory contents, which are created by associative linking. The frequency of occurrence has a role just as important as the evaluation of the importance of the information. A one-off event can be so formatively activating that partial aspects of the event already lead back to memory or even to the experience of the event, as is the case with post-traumatic stress syndrome (cf. Roth and Strüber 2014, p. 272).

Similarly, more frequently occurring information is related to other examples. In addition to the specialisation of areas which, for example, create a sensory characteristic in the sensory area because of the frequency of sensory stimuli (cf. Rösler 2011, p. 169), the brain also has a plasticity in its circuit structure that is supported by LTP and can express a higher frequency of use and thus of association. This connection can be simulated and verified in the (auto)associative networks. It is important to recognise that the interconnection structures can express a statistical accumulation which, although not directly conscious, can nevertheless lead to an influence on behaviour (cf. Rösler 2011, p. 228ff.). However, errors can also occur that creep in as the result of incorrect reference (newspaper reports) or little experience or distorted perception. In general, however, it can be shown that a large amount of information leads to the neural network, developing a correspondingly extensive background knowledge in this area with which good decisions can be made.

This realisation is the basis for the importance of the memory traces, because they depend on the number of stimuli and experiences. In addition, by shifting the amount of information, they can lead to changes being expressed. Traces of memory are the representation of associations and thus, together with the chunks, something that can be imagined as a cloud of words and meanings. They comprise a core of a meaning experienced by one person as well as the different associations emanating from this core depending on the situation; these support the cascade structure already mentioned in Sect. 7.1.2. The memory traces are the connecting lines between the terms, and it is not clear why they are there.

A larger and more extensive branching of the memory traces is limited by the fan effect, because a more extensive activation of several chunks is limited (cf. Anderson 1996a, p. 201). This limitation should be in the working memory. Here the aspect of expertise and intelligence already mentioned in Chap. 3 (3.2.2.6, Expertise) comes into play, which should extend this limitation accordingly. As the example of the chess players has shown, the expertise leads to a better processing of the information by merging it into more abstract chunks containing more information, which can then be better processed in the working memory.

A similar ability is shown for intelligence in terms of abstraction ability. One study showed that the use of longer words correlates with higher intelligence (cf. Lewis and Frank 2016, abstract). Combined with the previous information, this means that longer words are nothing more than superchunks, that is, several pieces of information combined into one, a particularly content-rich abstract of information. This accomplishment leads to a lower need for performance, because many are combined with one piece of information. The same can be said with regard to meaning-encompassing reading if a text can be better processed into overall information (cf. Anderson 1996a, pp. 402f., 437).

Traces of memory are thus the connections between the contents related to their frequency of representation as well as evaluated relationship. From this it can be concluded that the creation of something new can begin with this aspect and thus lead to a shift. This event can happen both through the conscious activation of less easily activated memory traces and through the connection of areas where no memory traces were present.

7.1.4 Production Rules

The production rules are the contents of the procedural memory experienced through the change of chunks from the environment and thus learned by them (cf. Anderson 1996b, p. 355).

As are chunks, production rules are thus a product of environmental incentives. These stimuli lead through the change to the fact that certain relationship structures express themselves among themselves. Because these are created by the relationship to the chunks, the production rules are also defined by the environment and the information content; they can also be defined as context dependent. Analogous to the chunks, the probability of retrieval predisposed by context dependency is also found here.

By merging production rules in the "production compilation," several production rules can be created to a new rule, whereby a similar picture is produced here in relation to a step-like structure. Of essential importance in the production rules are the emotions, which in this context are an additional factor in the process and are also activated as additional information in the chunks. Based on their own experiences and colourings, these also lead to specific patterns of action.

Thus, production rules can be used to create something new by changing the context for specific actions, as would be the case in the course of the changed action in a specific context. In addition, changing the emotional coloration is a part that can lead to a new result.

7.1.5 Hierarchical Control

The system of hierarchical control expresses in itself by the name of the control that something new is not possible. This impression is misleading, because it is not a question of control per se, but of the negotiation of different conditions or decisions: this happens at all levels. Therefore, it is a special form of production rule, but one that deserves to be looked at more closely because of its peculiarities.

Thus, the parts involved in the negotiation of competing parts are interconnected in such a way that the arriving signals activate their areas and inhibit the competing ones. This functioning leads to an overall mutual interaction, which, without a special weighting factor, leads to an increase in the contrast of the signal strengths arriving in both ranges until an on/off contrast is achieved, that is, the stronger signal in the input has prevailed over the other. In this constellation, those signals would be preferred that are easier to form, that is, those that have already experienced a higher or more mature degree of long-term potentiation (LTP). Because LTP is only formed through use, those signals would prevail that have already occurred more frequently and are more likely to do so.

The co-activation of associative areas, which creates a context reference both in the chunks and in the production rules, has a positive effect on this area, because the content that is probably most appropriate in the situation is accessed in this way.

Nevertheless, a purely experience-based settlement of the conflict would probably not be optimal. Therefore, several cascades of decisions are made in parallel on such a level down to the lowest level. These cascades are active in the decision coming from bottom to top related to the interpretation of contents, for example, in peripheral vision to differentiate between friend and foe (cf. Fademrecht et al. 2016, abstract and conclusion) or in the differentiation of noisy stimuli that represented either face or house (cf. Rösler 2011. p. 242). These connections are also carried out downwards in the higher areas to evaluate situations both emotionally and in terms of reward or punishment expectation. These processes have an impact on all areas and can thus ensure that a decision in a conflict area is not made solely on the basis of existing experience in this area, but also with the help of those from other areas that act as weighting factors.

As an example, an unclear stimulus in the sense of a friend/foe evaluation should be assessed in peripheral vision to make survival in a volatile environment more likely. One experiment showed that the recognition of a person's gestures in the distant field of vision in general gestures as well as the presumed intention of the same could be correctly assigned far above the level of chance (cf. Fademrecht et al. 2016, abstract and conclusion). Despite focusing on a central point, these processes thus lead to the presumed intention of peripheral movement sequences asserting themselves against the current focus in the recognition performance and can thus lead to a possibly necessary reaction without including higher areas at the same time. As this achievement, as also presented in the study, is of essential importance for social survival, the conflict regulations intervene in such a way that the differentiation of the evaluation of the stimuli is made possible.

This example is one of the bottom-up processes that interacts with a top-down process. For example, the rapid movements coming from the eye normally lead to a saccade in this area that enables focusing and better recognition. Parallel to the N100 and N400 shafts, a "known–unknown" as well as an "important–unimportant" assessment is processing, which results are that by shifting focus and including attention, these things become conscious. At this point, however, any necessary reaction has already been triggered and carried out. The areas to which the information was later assigned during the processing only then intervene, but in this regulatory process they shift the weightings and can top-down again.

For all these processes, however, there are limits to the change in weighting. In this case, too, the recurring trauma to be experienced should serve as an example, which consciously does not take place, but is nevertheless able to assert itself with the respective crossing of the activation thresholds.

Nevertheless, other elements can be mentioned in the hierarchy of control that produce different influences on the process of enforcing competing activity patterns.

Although cognitive control covers several aspects, such as setting goals, planning action to achieve them, setting priorities, and impulse control, only two should be emphasised here: attention as a regulator as well as a border to consciousness and context-related goal-oriented action.

7.1.5.1 Context-Related Target-Oriented Action

This element of cognitive control, which is defined by scripts, is, as are all elements of hierarchical control, a special production rule. The same principles therefore apply here. Nevertheless, this is of particular interest in the question of negotiating competing activation strands, as it is strongly schematically oriented by the scripts and strongly influenced by the order as well as the content.

The context-related goal-oriented action is created by frequent experiences of an active or passive nature. The framework of what is going on around the actual events is of particular character, as this can lead to a special role behaviour, as well as to different forms of the activity of a person. The decisive factor here is that the framework conditions produce certain behaviours and thus predestine them. The framework conditions can be the environment as well as specific organisational structures or people (constellations).

An example of this is the different behaviour of a person in a company with a strong and strict hierarchy, in which a person who is otherwise very creative in his/her free time always falls into an inhibitory behaviour. A special script can be film evenings, cinema visits, or common dinners, which partly differ between friends and a new romantic meeting in the operational sequences and contents.

This context-relatedness is also likely to play a role in automated behaviours, which is why they always come to the fore when it comes to enforcement under certain conditions.

To create something new here, either a new context would be conceivable or, although more complex and probably more difficult to realise, the creation of a new script or a new role. As a last point, one could also change something at the goal as orientation because it is about goal-oriented action. The question is which part of the reward system supports the goal and what kind of reward this would be.

7.1.5.2 Attention Control Center

In hierarchical control, attention can be seen as a special control performance because it can be targeted at all levels. This definition has been shown in the inhibition of saccades, as well as in the focusing of stimuli to be observed or in the recognition of more abstract forms in a search task (cf. Rösler 2011, pp. 95, 99f.).

The attention in itself therefore represents a special part of the control system consisting of many factors, which makes it possible to increase the focus in one area accordingly and to attenuate it in others at the same time. The attention thus assumes the role of a balancing mediator or a burning lens.

A further aspect of attention in its most extreme form may be the encapsulation from the environment, which enables the blocking of almost all surrounding stimuli for intensive introspection. In this way, the incoming stimuli are absorbed on the lower levels, but they do not penetrate the consciousness.

As already explained in Chap. 6 (6.1.2.5, Consciousness), consciousness itself is closely associated with attention, which is why the author does not make this distinction in a precise manner, as this would not be appropriate for this work, but a distinction would very probably be made.[3]

The consciousness is the part of the thoughts that is focused by the attention, which treats them accordingly, and differs from the preconscious in that this includes all other processes that have not yet crossed the consciousness threshold, but these can enter the consciousness. As an area of the unconscious, those parts of neuronal processing can be named that show activities but cannot enter consciousness.

This distinction is important because it expresses that attention can focus but has limitations. It also shows that there are activities that push into consciousness or attention focus, but which have not (yet) managed to do so.

This concept opens up the possibility of creating something new in this area by focusing attention on a certain area that was not previously in focus and reinforcing it. An activity pattern can also cross the threshold to consciousness from the preconsciousness through a different kind of activation and thus present itself as a priority thought. Furthermore, through intensive introspection, encapsulation from the environment can lead to previous activity patterns being reduced as an influence. The last option would be to reduce a restriction by losing attention and thus renegotiate competing systems of relatively equal strength. A special case of this omission of attention could be the free-tailing of attention, if it glides over different areas without dwelling. This form would presumably involve a short-term increase in the activity patterns, which could manifest themselves in the enforcement of new information and information combinations.

7.1.6 Neuroanatomical Approaches

Chapter 3 (3.3.1, Anatomical Basics) has already covered the basics of the cross-linking structure. From these, approaches can be derived to the effect that mutual interconnection has an influence on (information) density. The new brain cells formed in the course of neurogenesis are not only integrated into the existing structure, but also become more information rich with the increasing complexity of the interconnection mechanisms, both inhibitory and excitatory. This fact can be seen both in the place of cells necessary for orientation in the hippocampus, where at the same time the flexibility seems to suffer from integrating something new (cf. Maguire et al. 2006, pp. 1091, 1098–1099), as well as in the compression in the somatomotor area of musicians, which can lead to the development of the so-called

[3]At this point a work from Watanabe et al. titled "Attention but Not Awareness Modulates the BOLD Signal in Human V1 During Binocular Suppression" of 2011 should be mentioned, which was able through a complex design to determine different activity patterns in fMRI records, which were associated either to attention or to consciousness. However, also with this study there are diverging opinions leading to the body–soul dualism, which, as already stated, would go beyond the scope of this book. For the work in hand it is adequate to focus on attention and to distinguish between conscious and preconscious processing activities.

musician's dystonia. This is a degree of networking that goes beyond one's own area of specific representation, which leads to information distortions and allocation errors and thus no longer makes it possible to move fingers independently of one another, for example. This ability is only possible again through targeted reduction training, which achieves a solution by reducing the structures set up (cf. Rösler 2011, p. 159).

These changes in networking are achieved by two essential factors. The first factor is adult neurogenesis, which creates the integration of new ganglia into the existing network and can thus create new pathways between previously unconnected areas. Although no exact knowledge about the shifted or integrated information processing is yet known, it can be assumed that not only lost cells are replaced, but that changes in the network structure take place both in the information processing and in the information storage. As an example that reinforces this assumption, findings from experiments with mice are mentioned here. On the one hand, neurogenesis in mice changes in an enriched environment compared to normally growing mice with regard to the quantity of surviving neurons (cf. Kempermann et al. 2010, abstract). On the other hand, by experiencing something new, the same processing is repeated during sleep. This pattern is shown in an experiment by Ji and Wilson, in which these rats ran through a labyrinth. They recorded the cell activities of hippocampal as well as cortical target cells while the rat moved through the labyrinth. In a subsequent sleep phase ("slow wave"), the activities taking place there were also recorded in the target cells. An analysis using statistical methods showed that these sequences were replayed at shortened intervals. The impulse came from the cortex and initiated the same sequence that was active while actually passing through the labyrinth (cf. Ji and Wilson 2007, passim). Combined, these two findings showed both that learning increases the integration of new cells, which can be inferred from the increased survival rate of the new cells, and that the repetition of what has been learned during sleep is induced to reproduce these processes.

The integration of new cells can theoretically shorten a connection between unconnected areas and thus relate networks to each other in which such a connection was not yet present in this intensity. These new linkages would be able to affect both areas that are narrowly delimited locally and areas that connect over greater distances. However, because a too-unstable connection could lead to problems for the more distant areas because of the more variable excitation patterns in essential integrations of activity patterns, it can be assumed that a high ability to adapt could sometimes affect the mental health of the persons[4] involved.

These new neurons also exhibit increased synaptic plasticity compared to others, which enables them to integrate more quickly and makes them more adaptive. This synaptic plasticity, which manifests itself as LTP or LTD, is the second approach to

[4]The hereby mentioned aspects of mental lability are shown in different studies with creative people, who show a higher chance of mental health issues; these seem to be partially of genetic origin, as mental diseases are more frequent in families of creatives (cf. Andreasen 2005, p. 138f.). To what extent these are linked to the genetic predisposition of survivability of newly created neurons (cf. Kempermann et al. 2006, p. 781) is an interesting, but still open, question at this point.

changing the cross-linking alteration. The LTP as well as the LTD essentially change the excitability of a synaptic connection; these become more or less effective through this adaptation through incoming signals. Because such an adaptation can have an effect on both exciting and inhibiting transmissions, just as it is created by the additional training of boutons in the fine tuning already mentioned, the connection between areas that were not yet so well connected can be intensified. This step is done by either inhibiting or reducing the strength of other strong compounds, or by strengthening or reducing the strength of the weak compound. These diverse possibilities also make the existing networks more extensive and differentiated, which can also be used to create new ones.

In summary, it can be stated for the neuroanatomical aspects that the creation of something new can express itself both in the change of the synapses and in the change of the network, namely, through the integration of neurons. These two variants still contain differences in inhibition and excitation, just as they can also take place in conjunction with each other. These changing and complex interconnection possibilities allow both fast and fine-tuned adjustments in the neural network. Thus, these processes are stable for the creation of new things on the one hand and can contain complex information, but on the other hand they can also react to significantly changed conditions with a slight delay. The distance between the areas connected by new neurons is likely to have an effect on psychological stability from a critical threshold of ease of occurrence, which would accordingly become visible in creative people.

Because the crossing of the threshold of consciousness is mentioned as an essential aspect within the framework of the definition of thought, the neuroanatomical foundations for the emergence of something new in this model must be taken into account. The new that emerges is initiated by the changes on the cellular interconnection structure, but this does not yet penetrate the consciousness and thus the benefit of the new is compellingly given. It is only through a sustainable path or increase in specific excitation and activity patterns that the new is created by exceeding the not exactly defined threshold of consciousness. Novel reflexes or motor reactions are treated in this context in such a way that they penetrate the consciousness, at least in retrospect of the implementation, and have thus also crossed this threshold of consciousness. In all other cases in which something does not penetrate the consciousness, this limitation cannot be used to speak of creativity or something new in the narrower sense of this work, which is why it falls out of consideration.

The summary of the neuroanatomical approaches for the emergence of new qualities in the brain is very closely linked to the emergence of expertise, which is characterised by the fact that a specific area is very well mastered. In this context, expertise is understood to mean that the network patterns created for a complex of topics are very tightly interconnected, so that they can be easily activated as well as more abstractly. Therefore, expertise can be shown as follows: it is a dense and information-containing interconnection of the neuroanatomical level, which represents itself in a specific context with a circuit lying on a lower excitation threshold. This position makes it much easier both to connect to existing information and to connect with each other, and thus predisposes the artist to create something new.

In this context, another effect of the neuronal properties shall be considered, namely, the neuronal spontaneous discharge. The neuronal spontaneous discharge, which for a long time remained relatively unnoticed as background noise, is of importance in development according to recent findings (cf. Häusser et al. 2004, p. 9218). For example, it was found that already during cell development the spontaneous release of transmitter substances develops into a change in the formation of dendrite arms, which can also lead to targeted growth over greater distances (cf. Andreae and Burrone 2015, abstract and discussion). An in-depth examination of this phenomenon leads to the consideration that these spontaneous discharges have a role in the regulation of the flow of information within microcircuits (cf. Häusser et al. 2004, p. 9218).

The author sees this finding together with the medical fact that the resting membrane potential is caused by a lack of energy or by the inhibition of Na^+–K^+-ATPase (cf. Silbernagl and Despopoulos 2003, p. 44), in the generation of spontaneous discharges as a possibility that potentially well networked regions or microcircuits can experience a randomly initiated activation. This random activation is irrelevant to the extent that it does not in itself have any real function. In terms of creativity, however, it is conceivable that such a random discharge could create a connection with currently active areas that would otherwise not be associated with each other. A possible example of this difference is Guttmann's experiment, in which physically identical tones were played to test persons constantly for a longer period of time. The test persons took these tones as partially different, which was documented by means of a push-button (cf. Guttmann 2016, p. 1f.). This experiment, which already served as an aspect of confirming the constructivist approach in Chap. 6 (6.1.1, Constructs), showed that this different perception was also reflected in neuronal activity. With the understanding of the just mentioned energy deficiency or the inhibition of Na^+–K^+-ATPase, this could be a kind of "fatigue effect" that could be expressed by the same repetitive monotonous external excitation in a delayed internal activation. Although this is not very likely, as neurons can trigger an action potential up to 1000 times per second, just as they allow only about 1 in 100,000 of the intracellular ion quantity to flow through the membrane in an AP (cf. Silbernagl and Despopoulos 2003, p. 43), a deviation in accuracy caused by a short-term temporary fatigue of individual cells or associations would nevertheless explain such a shift. In a simple final calculation, the total intracellular ion quantity would flow through the membrane in 100 seconds in an intensively used neuron cell, without taking into account the constant restoration of the resting membrane potential. Combined with the findings of Bartol and colleagues that the fineness of the gradation of synaptic transmissions can be fine-tuned many times more than previously assumed, this time would probably be reduced even further. It has not yet been included in this entire process that the continuous activation is primarily oriented towards and depends on the transport materials necessary for the return transport, which require the energy supplier ATP for this transport.

A connection to this area can be found in the study by Silver et al., who showed in an experiment under laboratory conditions that neurons and astrocytes react differently to energy deprivation but are dependent on a complex function of the internal ATP production capacity to maintain the ion balance (cf. Silver et al. 1997, abstract).

The conclusion from this is followed up in a later part of the chapter, but it shows that chance could also be a certain component, not only in experiencing and interpreting, but also on the cellular level and thus in its effect on the new.

7.1.7 Recapitulation

In summary, it can now be deduced from the existing model that the new can be created by various elements. As a part of the essence it seems to be the realisation that the brain orients itself in principle to the stimuli of the environment, but does not represent the stimuli purely constructively, but associates these and connects them with each other. This connection is already something new in itself, which is why the author can subscribe to Carson's idea that the brain is above all an organ of creativity (cf. Carson 2010, p. 39). These associations are the enrichment of sensory information with a relatively close other sensory information and thus an extension of the information, whereby proximity does not only mean time or space.

For the creation of something new, the following aspects have now been derived from the model and its links:

- Change of stimulus, which can be expressed specifically in the type of stimulus as well as in the change of the environment and thus the context. The shifting of stimuli changes the associations and thus the processing.
- The level reference, which is used for the interpretation in the processing of information units. For example, in text analysis one can distinguish between sounds, syllables, words, and sentences as well as texts themselves. An analogous cascade of distinctions is also applicable to other senses and their processing modes.
- Traces of memory, in that these elements, prepared with a preactivation, connect with others or allow them to be transferred into a broader field and become more diverse. This change is possible both by paying attention and by reducing control mechanisms. Thus, the association routine becomes breakable and can create something new.
- Use of rare information by deliberately activating it. The presence of rarely used information is also expressed in the low activation strength of these elements, as they are linked to few contexts and information. By turning to these, it is possible to break through an association routine, analogous to the traces of memory.
- Production rules, if they are applied in a context other than the permanent context. By combining production rules with a specific environment from which they have been derived, new things can be created both by reinterpreting and linking the environment and the production rule.
- Emotional colouring, as special context information for production rules. The emotional colouring of a situation can also lead to a different contextualisation,

and vice versa a contextualisation can contain different emotional colourings. By attempting a consciously different activation of emotional framework conditions, regardless of whether this is the case through mental techniques or stimulation with technical aids,[5] the context information is changed in such a way that new connections and associations can be found.

- Attention as an inhibitory control mechanism. By focusing on one aspect, attention can suppress other possibilities, which might be similarly strong in activation, by contrast. In this case, reducing attention would be beneficial so that alternative options can be allowed and brought to the fore.
- Attention as a strengthening control mechanism. With the help of contrast, the attention can stimulate a strengthening of weak signals by consciously focusing on different areas, which can then occur as new signals crossing the threshold of consciousness. It would have to support a conscious system of strangification/thought oscillation to be able to integrate distant areas as well.
- Intensive introspection as a particularly pronounced aspect of attention focusing. The intensive introspection separates the thoughts from the external stimuli through a strong focus and thus enables a decoupling of context information and activations. In this form, concentrated on purely internal processing and information routines without external influence, the changes also enable the creation of something new.
- Reduction of control mechanisms when these lead to lateral disinhibition. By reducing the control mechanisms, different competing systems compete directly and can, without influencing higher areas, lead to new or different activity patterns becoming established and leading to different results.
- Neurogenesis as an aspect of changing neuronal networks. Neurogenesis can be increased by stimulating production through exercise and by increasing the likelihood of survival of new neurons through learning new information. In both cases, the integration of new cells enables a changing network structure that can create something new from them.
- LTP/LTD as an aspect of changing the energy efficiency and thus the excitability of individual neurons, which change the surrounding network through activation: this can happen through stimulation as well as spontaneity. In the latter, the effect of monotony could be an important aspect that can be addressed.
- Expertise as an efficiently interconnected network structure. Because of the existence of much information and its interconnections, an easier excitability is possible through expertise, which expresses itself similarly to the variations

[5] A few options are mentioned here in short. Mental (training) techniques include exemplarily stress relief methods, Begriffsentsressung (literally translated term or word relaxation), and mental imagery training or moment of happiness techniques (cf. Täuber 2016). Technical aids could be neurofeedback training as well as a DC-stimulated device for a relaxed sleep in PTSD-patients in the US. Both aspects enable a change of emotional activation and association. Additionally, psychotherapy could be mentioned here, the goal of which is an alteration of emotional states to increase health and life quality.

mentioned in the context of the random triggering of action potentials. The same is true of the fact that it is based on easier excitability, which enables new and nonresident information to be better linked to existing information; this also creates something new and expands the field of expertise.
- Chance, as it takes place in its alteration of neuronal networks and interconnections in the two points mentioned, expertise as well as LTP/LTD. This association should not only be considered as part of neuronal spontaneous activity, but also in the sense that parts of the neuronal network can change randomly from unspecific and undirected influences. This change is conceivable in both a positive and negative direction, as, for example, in the context of strokes or other brain damage both restrictions can occur, and this can lead to the development of special talents and abilities, not to say savant syndromes.[6]

These derivatives from the model are now related to different experiments in the next sections.

7.2 Experimental Reference Setting

In this chapter, the aspects considered to influence the creation of something new are now related to various investigations and experiments to offer an approach for the explanation of investigation results. This inductive approach should allow a confirmatory or deprecative insight to be gained. The experiments were selected on the basis that they could be related to the creation of something new or to creativity research.

7.2.1 Language as Context Information

In one experiment it was shown that language influences an essential part of context information, resulting in fundamentally new views and interpretations for a specific situation.

[6] The loss of speaking abilities or partial paralysis in stroke patients is well known, as may the gain of extraordinary well-developed abilities as in Jason Padgett, who became a math savant after a massive blow to his head. Other patients examined by Darold Treffert experienced similarly special changes in their lives, such as Derek Amato, who became a pianist without years of training, by hitting the ground with his head after a jump (cf. Piore 2013, passim).

7.2.1.1 Experiment Description[7]

In this experiment, subjects were shown 6-second-long short films in which an action was performed that was goal oriented, such as a woman approaching a vehicle. As the English grammar of the present progressive with the "-ing" form places a special focus on the execution of the action in the case of current actions, and because the objective of the action also takes a back seat to the action, the perception should differ from that of a language with a holistic approach, in which the end points are included in the consideration.

To verify this approach, a first experiment was conducted by Athanasopoulos, and colleagues showed the videos to two groups. These groups consisted of monolingual English or German speakers, whereby the monolingualism of the German speakers was given to the extent that they spoke poor or very poor English. In addition to the monolingual persons, randomly distributed English-speaking and German-speaking bilingual volunteers whose first language was German were added evenly. The respective groups were instructed only in the language of the monolingual participants.

For the videos, 19 triads were shown twice, which included a goal-oriented action without reaching the goal (target clip X, "motion towards a specific goal") and a movement with arrival (alternative A, "motion with arrival at a goal") and a movement with such a distant goal that it could not be named (Alternative B, "motion with a possible goal far away from the agent"). The test persons had to tell the investigator whether X was more similar to alternative A or alternative B.

In a second experiment, a follow-up study was conducted with the same materials, in which the subjects were to repeat three two-digit numbers determined before the first video clip was shown loudly during playback until the last clip of the triad was completed. They should then give the answer to the investigator. The difference in this experiment was mainly intended for another area. The subjects were all bilingual and their first mother tongue was German. The test persons were again divided into two groups, one of which communicated with the instructor as well as with all written material in German; the other group did everything in English. To interact with the language context, the two groups had to conduct the interference task during the first half of the experiment in the language of instructions, and the second in the other. At the same time the interference task language changed, the answer was to be given in the now changed language. This step ensured that in the course of the experiment both German and English could be seen as activated by all test persons for the numerical pronunciation on the one hand and for the interpretation of the screen clip on the other hand. To make the results in this experiment generally comparable by changing the interference task, monolingual subjects in both German and English were confronted with the same task; this also allowed the influence of counting on the results, if any, to be recorded.

---

[7] The experimental description as well as the results are from a study by Athanasopoulos and colleagues (cf. Athanasopoulos et al. 2015).

Although the overall tendency was towards the variant that performed the action, which, according to the authors of the study, was caused only by the achievement of the objective and not the completion of the action,[8] the results confirmed that the aforementioned assumption had been correct. The monolingual persons tended in the categorisation more towards where it was to be assumed from the language grammatically. Thus, the preference of the German-speaking persons for the X-Clip with the Action Completed (Alternative A) was set more often than that of the English-speaking persons. Furthermore, it turned out that the bilinguals, influenced by the respective environment, found themselves similarly pronounced in preference as was the case in the rest of the respective group. This preference resulted in the significantly higher categorisation in the direction of "completed action" in this order: monolingual German speakers (40%) before bilingual persons in German-speaking context (38%) before bilingual persons in English-speaking context (27%) before monolingual English speakers (24%). Thus, it became apparent that the classification of the observation of an excerpt of an action in relation to the completeness is significantly influenced by the contextualising language.

The second experiment showed that the general tendency to favour the still ongoing action was again present. Furthermore, it became clear that a change had taken place comparing the phase before the language change and after the language change. In contrast to the first experiment, the influence was reversed. Compared to the German-speaking phase, during the English-speaking phase more completed actions were chosen, which resulted in the following distribution:[9] Pre-switch and start English (English) with 41%, Post-switch and start German (English) with 36%, Post-switch and start English (German) with 33%, and Pre-switch and start German (German) with 27%. Here it becomes clear that the language of the reproduction of the numbers affected the order of probability of the selection of the completed action. Compared to the first experiment, this was the other way around. The experiments carried out with monolinguals provide information about this, as both of these are on a par with each other with a probability of about 33%. The study interprets these results in such a way that they show what was already known from other studies: the language currently used takes a back seat to a second language for parallel tasks. So, if English is spoken, then the presence of the ability to speak German leads to this being used for the task of interpreting the situation and thus, although English is spoken, the processing of the clip takes place under German influence.

The authors thus showed that the working language, as does verbal interference, produces a shift in the preferences of interpretations of situations. In addition, they proved that in bilingual persons, through verbal interference, the unused language

[8]As clarification, exemplarily reaching a door may be used. A person who reaches a door has already reached the goal, but the arrival does not finish the act of walking through it, which probably influences the setting according to the implicit script. Therefore, the goal is reached, but the activity (sequence) is not yet finished.

[9]The languages in brackets are the languages used to shout the numbers out loudly as well as these were the languages in use for communicating the answers to the investigators.

expressed itself in the categorisation patterns; this also means that the noninterfered language is used in the course of linguistic encoding through the immanent conceptual content.

One aspect that is not mentioned here and that could still be mentioned is the fact that both the German and the English monolingual test persons have experienced a shift in comparison with experimental study 1, namely, to the effect that the English persons increasingly chose the categorisation in the conclusion of action just as the German persons reduced it. The influence of the execution of the playback task has already caused a shift in the interpretation of the situation.

7.2.1.2 Analysis of the Experiment

The experiment focuses on the language context and shows that different results can be obtained in the interpretation of an event. The influence of the language either creates an access that is more in the seclusion or more in the still-ongoing implementation of an action. This shift is based on various mechanisms. From the perspective of the authors of the study, the grammatical representation of an action has a focusing effect on the current action through its reference to presence. Or rather, through their detachment from the implementation, rather holistic approaches will come to bear.

Consequently, this means that a change of language can bring about a change of perspective. Such a change of perspective would allow a new interpretation to emerge from one and the same situation. This new interpretation can be as much a part of the ordinary creativity as it can be an impulse for an extraordinary one. But what could be the reasons for this?

At first, it should be reminded of the chunks. Chunks are the information units. They are independent from the respective level as well as they are from the context. From these chunks outgoing connections are put over memory traces to other information units. The so-called isa-pointers frame these more detailed definitions. If one includes in this case the fact that the languages mastered at a very high level are activated in similar areas,[10] then it becomes clear that the effects are quite obvious here. If one sees concepts in this context as something experienced from the environment, which is defined by this experience with the associated properties, then the understanding of the constructed associative representation becomes clear: this is chance and limitation at the same time. Through the acoustic association of a term

[10] In a PET-study, Perani and colleagues had already showed in 1998 that different activation patterns differing a native language (L1) and a second learned language (L2) are not primarily determined by age but the level reached (cf. Perani et al. 1998, passim). Hirsch and colleagues detected that (foreign) language-relevant networks are found in the Broca and Wernicke areas, but differ referring to proximity. So, people raised bilingually have huge overlapping use of the area and therefore very close center points of the different languages. People who learned a high level of a second language later in life have center points with a greater distance in between these points. Additionally, their overlapping area of use was smaller (cf. Hirsch et al. 1997, passim).

such as "cube," it is associated with the visual or tactile stimulus that is relatively close to time and space. By a corresponding frequency of the meeting the connections interconnected in the neuronal network are strengthened and are expressed in the terminology, which also includes other information. These data are certain information derived from the visual as well as the tactile sources, which lead to the fact that there is a certain basic concept of this term. As already described in the development of chunks, they express a certain environmental information content. Thus, they also contain the relationship to the grammatical connection, which is learnt through the acoustic absorption of stimuli through the frequency of experience and is set in relation to this. Thus, a term includes not only clear information about its nature, but also about its relation to others. This part of the chunk, which can be depicted under the term production rules, is thus also language- and grammar related. This addition to the terminology also reflects the interpretation patterns of the language, which are holistic or focused, as well as cultural aspects, for example, individual or collective orientation (cf. Brozinsky-Schwabe 2011, p. 93).

At this point, a short excursus should make this part of the theory more descriptive, because it is essential for understanding: A chunk can be imagined in such a way that it resembles a "word cloud." This structure is dense in the centre and becomes increasingly brighter and less precise with increasing distance. By activating a specific context, the density of this cloud is shifted to this area, where it can expand further. Because clouds can also be charged, propagation by the voltage differences depending on the orientation can be imagined here. By shifting through the context, the orientation in the context is conceivable, which at the same time also moves the linguistic shift. For persons bilingual in German and English, the term "church" also exists for the term "Kirche:" these will be similar in their essential characteristics. With the grammatical contextualisation of language, however, different interpretations of the same term can be triggered here.

This realisation makes the relationship shifts understandable. When an action is spoken of in a process, both the chunks and the associated production rules are activated. Through special relationship structures from the environment, special concepts are also activated. These for event types as scripts (cf. Anderson 1996a. p. 157) called chunks contain the experienced references and also represent a kind of "event cloud" through this reference. Do not forget at this point that the chunks offer the advantage of depicting all kinds of levels of representation and are so abstract as a basic unit of information that they enable a unit independent of levels that simplifies the thinking of cognitive processes. This is something to think about, because an action such as moving towards a vehicle could be related to influencing scripts. But as the chunks as well as the rules of production are to be thought of as part of the experienced and related reality, the grammatical structure of the relationship is to be seen as a special rule of production. Production rules arise from the fact that these specify changes in chunks (cf. Anderson 1996b, p. 357). Consequently, this means that these production rules exist as a grammatical appendix to the chunks: in this case it does not matter whether they are interpreted as the scripts or as the words occurring in this process in the respective language coding. In this way, the differences in preferences assumed in the introduction to the study on current or targeted action and thus the action including the goal in this part of the knowledge become understandable.

If this fact is now taken further, it becomes clear that only one of the possible production rules can be expressed and applied in the working memory, as is the case with the chunks. Because in experiment 1 preference for bilinguals shows that the language which is more coherent in the context prevails, that is, the German language in the German environment and the English language in the English environment, this confirms the dissolution of competition assumed by Anderson by the context reference, which he described in the ACT-R as contextual priming (cf. Anderson 1996b, p. 360). In addition, in the Integrated Theory of Mind, Anderson brings the importance of the chunk in the past and the topicality reference introduced by the associations in the context into a formula through a closer definition (cf. Anderson et al. 2004, p. 1042). Because only one of the two languages can assert itself in the reference setting, this is also applied in the evaluation in its world view, which is expressed in the significantly different evaluation of the closeness of the action.

The language change test in experiment 2 also shows that the working memory theory, which deals with the processability of only one production rule, is confirmed here as well. What is interesting about this approach, however, is that the activities taking place on two different channels, the memorisation and repetition of numbers and the processing of visual information, indirectly influence each other and thus cause the shift in tendencies. In this context, Baddley's as "slave-system" labelled aspects of the working memory are being activated: the phonological loop, which is responsible for processing acoustic information, and the spatially visual notepad, which serves to process this information (cf. Baddeley 2003, introduction). This also explains why the speech task, which is based on the phonological loop, is hardly disturbed by the visual task. At the same time, it becomes more understandable that the language used in the speech area by the speech task is blocked for visual processing, which is why it seems to resort to an alternative free language. As a result, because interference takes place on a different channel, it will lead to the categorisation of the other language asserting itself. This validity becomes striking insofar as the categorisation only takes place after the numerical task has been completed. Thus, the visually processed information, which has activated even more active patterns of the alternative language and thus its interpretation models in the memory traces, is processed in language two.

Following this understanding, the explanatory approach could also be applied to the monolingual test subjects, which showed the shift in categorisation as a completed action at the 33% level for both language groups. As in this case both groups of persons were restricted in the processing of the visual information, the action control guided by the central executive could interfere with the execution of a parallel goal-oriented action, counting, to such an extent that this could be expressed in a one-third probability. Because this was the case for both language groups to the same extent, a language-independent aspect should be responsible. The central executive would be one, especially as it is thought to be the element responsible for attention control. Because of the absence of a "second language buffer," compensation would be omitted in this case and the language component would also be eliminated as a variable. Although this would have to be verified in separate experiments, it could certainly offer an interesting approach. Among other things, further consideration would have to be given to whether and to what extent the seclusion or the still running of counting has an effect as contextual priming in the sense of Anderson and whether this could refer to special performances.

From these derivations, however, the interesting thing for the research-guiding question is how something new can emerge here. Because only what and why various elements in this experiment could have had an effect on the results has been mentioned so far, now the reference to the creation of something new is to be established.

As mentioned at the beginning of this chapter, a change of language can lead to a shift in the interpretation of the environment as well as of events, situations, or data. If now a different view can be produced by a change of the language, which must be mastered on a high level, then with this view also shifts in the "word cloud" are possible, that is, the relative focus of a term, a situation, or the like comes by the change of the language to a new reference. In this sense, this shift would activate alternative traces of memory and thus take other paths.

As an approach, the conscious use of an alternative language in another context can be derived from this study, just as the monotonous repetition of non-topic information in different languages enables activation through visual stimuli in the other language area. This activation in the less frequently used areas bypasses the otherwise effective control mechanisms, which in routine evaluations enable the inhibition of information that may already tend to be present. Because this takes place consciously under attentive action, the probability also increases that in this situation information that is still too weak can assert itself against other activity patterns as a result of the lack of competitors as well as the contrasting patterns that support attention in such a way that this exceeds the threshold of consciousness and is thus temporarily actively perceived.

This is an experiment that causes the shift of the context and thus an influence on the activated chunks and production rules and with them their traces of memory. In addition, attention here has two important roles: by turning away from normal patterns or by blocking them within the framework of monotonous counting, these patterns step into the background and at the same time, by turning to alternative patterns, they come to the fore compared to competing variants and can thus create something new.

7.2.2 Irritation as a Basis for Cognitive Flexibility Increase

In creativity research, cognitive flexibility is often seen as an important factor for the emergence of creativity and measures this also in various tests such as "unusual use task" by Guilford (cf. Ritter et al. 2012, p. 961). The purpose of this test is to find as many possible uses for an object as possible within a certain period of time.

This example makes the flexibility of thinking as well as the originality and fluidity of thinking visible and measurable to a certain extent.

7.2.2.1 Experiment Description[11]

Two experiments were carried out in this research.

The first experiment involved 61 students who were randomly divided into three groups. The first group ("active unexpected event") was exposed in a virtual reality to three unexpected events that ran counter to the physical experience.

- The first event was a suitcase standing on a table. If the test persons moved towards it, it became smaller, if they moved away, it became larger.
- As a second event, a change of the person's locomotion speed was recorded. Each step of the test persons was extended by a factor of 1.5, and thus the speed was changed.
- The third event was a toy car standing in the middle of a table and a bottle at the edge of the table. If the test persons moved towards it, the car began to roll towards the bottle and pushed it off the table. However, the bottle did not fall to the floor, but began to float slowly upwards as soon as it had left the table.

The second group ("active normal event") experienced the same events as the first group, but with the difference that everything behaved as if everything was normal. The suitcase became smaller when leaving and larger when approaching, the speed in the second event remained at the ratio 1:1, and the bottle of the third event fell to the floor.

For the third group ("vicarious unexpected event"), the events of group one were recorded and played to the test persons in a film.

Following the event exposure, the volunteers were each treated with a "Guilford Unusual Use Task" and then carried out a test with the PANAS of Watson, Clark, and Tellegen to record affects, states, and traits (cf. Watson et al. 1988, p. 1063) to exclude influences by these factors. The analysis of variance showed that group one had a higher cognitive flexibility than groups two and three. Furthermore, it turned out that the affect influences were not present, but only the active and unexpected events were the decisive factor.

To go further into these findings, a second experiment used a change in schemata as an unusual element. Eighty-one students were randomly assigned to four groups (first group; "active schema violation"; second group, "active schema normal"; third group, "vicarious schema violation"; fourth group, "vicarious schema normal"). The event was the making of a breakfast sandwich with butter and chocolate sprinkles, a popular breakfast in the Netherlands. The test persons of the active groups were given instructions about the activity to be performed on a screen and had to

[11] The experimental description as well as the results are from a study carried out by Ritter and colleagues (Ritter et al. 2012).

follow this procedure step by step, whereas in the indirect variants the test persons were again shown only a film in which an actor performed the actions.

For the violation groups, the sequence of events was changed so that the test persons first had to sprinkle the chocolate sprinkles on a plate, then spread the bread with butter, and finally place the bread with the butter side down on the plate with the chocolate sprinkles.

Under normal conditions, the usual order was followed, first placing the bread on the plate, then brushing the bread with butter, and finally sprinkling the sprinkles on it.

Following the events, a Navon task (cf. Navon 1977) was performed with the volunteers, as well as two "Unusual Use Tasks." To round off, the PANAS was again used to measure affects.

The results again showed that the subjects of the active and unexpected processes had performed better than all others in terms of mental flexibility. The values in the first "unusual use task" compared to the second one were higher in cognitive flexibility; this also applied to groups two to four.

Overall, after both experiments, it turned out that the persons actively acting in atypical changes for the normal environment experienced an increase in cognitive flexibility compared to inactive and only experiencing as well as actively acting persons in a normal environment, and thus more creative solutions could be found.

7.2.2.2 Analysis of the Experiment

These experimental setups focus on the distinction between normal and abnormal events. Thus, the usual and so far experienced experiences and abstractions are not illustrated but undergo an irritation. These irritations cause a significant change in cognitive flexibility, which leads to more extensive as well as more numerous and divergent responses in the Unusual Use Task. Through the irritations, it was seen that the creation of the new has been significantly increased. But how would the model of thinking here make it possible to explain this increase?

On the one hand, it should be noted that the events experienced by the test persons in the first experiment were of very frequent occurrence. Just as the speed of movement, which can change when viewed through life, but only very slowly and barely noticeably, is always connected with spatial orientation and the location in space as well as the possible change in position is experienced, the changes in perspective and proportions in this context are also everyday experiences. These two aspects relate very strongly to the combination of motor and visual processes and their interaction. In this way, the changes in depth effect and proportions in relation to motor processes such as walking are processed in parallel. The third event that something is knocked off a table is not a daily experience, but still a frequent one, and is also related to falling down from an early age. Gravity as such may not be known to all persons, but the effect of it certainly is. These are therefore three very frequently experienced events, experienced in different contexts and environments, of which it can be assumed that from the diversity a strongly abstract and simplified illustration with many characteristics already exists. The chunks are extensive and have many traces of memory, as also applies to the production rules.

On the other hand, it can be assumed that this frequent experience may lead to a certain amount of expertise, that is, that the networks experienced in these situations can enable a more rapid spread of action potentials and that the spread of excitement could therefore not only be possible in different areas, but could also be more easily stimulating.

Experienced situations are gone through in different ways. Experiment 1 distinguishes between active and passive and between normal and abnormal with regard to the experience of the irritation environment. This distinction is interesting in the area of active variants, because the events are experienced interactively here. As has already been shown experimentally, simulations are experienced similarly to reality and cannot be distinguished from real simulations on the basis of anatomical parameters such as heart rate (cf. Ylönen et al. 1997, abstract). It is also known that the memory of events leads to a relatively similar reactivation of the activity patterns of the event (cf. Rösler 2011, p. 198). In connection with the spread of the activation of associations along the networks connected with these (cf. Anderson 1996a, p. 183), that is, along the traces of memory, as well as the understanding of the existence of schemata, classes, and scripts (cf. Anderson 1996a, pp. 150f., 153f.–157f.), it can be stated that the representation in the virtual environment is equated in its experience with a real one and is thereby expected to activate the same brain regions. In this case, this applies in particular to normal processes and events.

Because the abnormal events deviate from the expected processes, these pathways are not followed. Normally, the frequently experienced connections in the course of an event and the resulting lowered excitation threshold in the synapses interacting for these processes suggest an easier spread of the action potential should be possible. This effect, also known as priming, also expresses the expectation and preparation in a situation, which results from the context, but also from the experience. This event activates the expected memory traces for the respective scripts or events from the chunks. Because these do not occur, however, but proceed differently, an irritation occurs, which has thus activated or prepared the activity pattern to be expected, but experiences a different one through sensory perception. These different activity patterns would lead to the expectation that more and more different areas would be activated in the brain than would happen in the case of normal experience.

In this case, the experience is extended and the memory is also extended, because the two are linked by similar areas. The motor sensory connections run together in convergence areas, which are responsible for the connection of this information. Expectations, however, mean that these areas are already preactivated and, corresponding to priming, are likely to be in a state of easier excitability. Because here the excitement fails at least to the extent expected and is expressed in a different way by different perceptions, a different pattern of activity develops; this generates an interference with the previous states of knowledge and leads to a different propagation structure. However, because both activity patterns were used, an increased excitability is to be expected for an activation soon after. This increased activation would explain how the test persons can achieve a higher value in the creativity test in the extraordinary test sequences. However, the question of the distinction between active and passive remains open.

In motor active and passive motion sequences, the same activity patterns are activated in the contralateral sensomotoric areas. The same happens in the supplementary motor cortex involved in learning sequences of action. However, a difference can be observed in the basal ganglia and the gyrus cinguli, which are only activated during active movements (cf. Weiller et al. 1996, pp. 105, 109). From this insight it can be deduced that the routines in the basal ganglia are also more activated in the action in the virtual environment than in the pure viewing of a film. Although some studies show that passive behaviour over time and the content learned can have an effect on behaviour (cf. Spitzer 2009, pp. 366–372), we do not consider the effect of learning in this study, but that of cognitive flexibility. The insight of Weiller and colleagues is quite interesting for the latter, as it shows that more areas are activated in the action itself than in the passive posture. Together with the previous paragraphs in the differentiation between the active and the passive variant in the experiment, this suggests that the areas active for the expectations and preparations are not activated in the passive variants. The passive watching of the film should, analogously, only produce the afferent activations that would not have an effect from the basal ganglia back into the retrieval of the plot routines, at least not with the same intensity and impact. The difference thus explained between the active and passive variants further leads to the question of how this irritation can have additional effects.

As an important effect the attention was mentioned, which can lead with its focus to the reinforcement of contrast effects. Because another factor occurs in this situation, namely, the use of rare information—a perception that has never been experienced in reality but can be seen as such—leads to the activation of other brain areas. This activation can result from the experience from movies or news (think of the movie "matrix" or a report about astronauts in space or about a parabolic flight), which could be activated here as association patterns, or simply be perceived and irritated completely without approach, because not a single experience has yet produced a visual enlargement of an object as it moves away. Despite their novelty or rarity, these alternative excitement patterns link up with other networks: this creates more extensive activations, which could have an aftereffect immediately following execution of creativity tests.

In connection with this interesting aspect, the question arises as to why in the second experiment the performance at the first "Unusual Use Task" was better than the second. This difference is interesting because the memory traces are assumed to have a longer influence on other activations: these can thus have an even longer disturbing effect (cf. Anderson 1996a, p. 393), whereby in experiments these influences are considered only over short times, because the necessary interferences are usually built into the next words or sentences. In this experiment, however, the question is whether the "Unusual Use Task" performance could be traced back to the fact that the distance in time created the variation, or whether the subject was more restrictive. In the first example, the task asked was "What's making a noise?" For the second task, it was required that the test persons generated different benefits for stones ("generate uses for a brick"). One could argue here both that the questions were differently specific and also that they might have been experienced in differently extensive contexts. This second part would speak for several chunk connections and networks and thus for a different creativity achievement.

If, however, the theory of the propagation disorder of memory traces, as achieved by priming, could also have longer effects, which is to be expected from the synaptic change in conductivity caused by short-term LTP already occurring after seconds, this would be an additional explanation for the different values of the two creativity tasks. The selection decisions necessary for the test would at the same time represent a limitation for the follow-up test, because the test would still be on other tracks, that is, focused on a certain aspect in the breadth of creative possibilities.

One aspect that can still be used for the increase in creativity in this experiment is the reduction of control mechanisms. Because a shift in reality has taken place in this experiment as a result of the irritations, the previous decision and behaviour patterns can only be subject to limited coherence. In terms of Anderson's calculation model, this means that the production rules have never had to operate in this context. These rules are in competition with each other and are usually negotiated in the course of the utility analysis, but here they are unrelated to each other, as is expressed in the production benefit equation (cf. Anderson et al. 2004, p. 1044):

$$U_i = P_i G - C_i$$

where P_i is the estimated probability of achieving the target using the production rule *i*, *G* is equal to the value of the current target, and C_i represents the estimated cost of applying *i*, typically calculated in time. P_i and C_i are learned by experience from the application of the production rule (cf. Anderson et al. 2004, p. 1044). Because there can be no clear values for this situation, however, these could lead to shifted representations of the associations and thus to completely different or broader application of production rules.

Because the various control mechanisms thus enter into a new relationship, shifts up to the avoidance of higher restrictive control elements are also possible here, which can lead to parallel processing on the lower levels and also create new connections or allow other networks to break through.

These findings apply to both the first and the second experiment of this research group, because both have led to a change in environmental processing from the irritation. This effect was confirmed both in the general complex situation and in a minor schema change/disruption.

This experiment thus confirms that a change in the environment, regardless of its size, can already produce a change in cognitive flexibility. By reference to the model of thinking, it was also possible to show why active involvement is likely to be important. It has also become clear on which mechanisms this increase in flexibility could be based.

7.2.3 *The Body as an Influence on Thinking*

The experiment examined in this chapter is concerned with the interaction between body and emotional mood, which appears in body language in the form of both gestures and facial expressions and thus establishes a relationship to the representation of environmental stimuli in activity patterns, as determined in Chap. 6 (6.1.3.1, The Brain as an Organ of Association Representing Constructions).

Body language itself is associated with the physical expression of mental states, which are always to be viewed in context but can be indicators of mood and thoughts (cf. Pease and Pease 2004, p. 10f.). Thus, the mental state is reflected in the posture, movements of the body, or similar aspects. However, as the representational function, the association function of the brain as well as its alteration, were already addressed in Chap. 6 through the possible approaches identified in Sect. 7.1.7, a further insight could be gained into the alteration of body language.

This experiment considers in detail the influence of the contraction of the facial muscles during laughing or smiling, in particular with the muscle orbicularis oculi, which can only be consciously contracted with the pars medialis during smiling and only during a "genuine" smile that incorporates pars lateralis. In both types of smile, the zygomaticus major muscle is also involved (cf. Ekman and Davidson 1993, p. 342).

7.2.3.1 Experimental Description[12]

In this experiment the term "other smile" is used for the smile consciously produced in the previous paragraph without the pars lateralis and the designation "Duchenne's marker" is utilised for the previously smile described as real, thus including both the zygomaticus major and the pars lateralis and medialis of the orbicularis oculi in the smile.

In this experiment, test persons are examined by means of EEG measurements in the conscious execution of the two different smile variations, and the results are compared with activity patterns of emotional activating stimuli. After an individual examination, 32 of 45 students of an introductory psychology course were rated as good in the conscious control of various individual facial muscles relevant to the experiment and continued to be used for the experiment. Of these, 5 did not appear in the experiment and 12 others had to be excluded from the study after the occurrence of artefacts during the EEG recordings. One last participant was eliminated after an equipment error, leaving 14 people for the study.

The test persons sat alone in a room for the experiment after the EEG electrodes had been applied to them. They were told that a trainer was sitting in another room who would see their faces on a monitor and give further instructions via an intercom

[12] The experimental description as well as the results are from a study carried out by Ekman and Davidson (1993).

system. The experiment then began with the recording of a baseline measurement and was then carried out in nine rounds, two of which concerned the smile variants to be investigated. The instructions for the "Duchenne's marker" given to the test subjects were "raise your cheeks" and then "part your lips and let your lip corners come up." For the other smile, the first instruction to lift the cheeks was omitted. In both cases, once the target position was reached, the subjects had to hold it for 20 s. The sequence of the two variants was varied over the course of the experiment, and the entire procedure was recorded on the EEG.

The result showed that, as other studies on the emotional evaluation of a situation have already confirmed, the two variants of the smile caused in the anterior regions of the brain—the lateral frontal, the medial frontal, the anterior temporal, and central temporal—significant and almost significant differences in activation, but this was not shown anywhere in the posterior regions. Furthermore, it became apparent that there was also a strong correlation in activity between the smile type and the hemisphere. This correlation was expressed by the fact that the "Duchenne's marker" type smile led to a significantly higher activation in the left anterior temporal region in 12 of 14 subjects than was the case with the other smile. This distribution was found in the studies mentioned at the beginning of this paragraph in the same way for distinguishing positive feelings in the evaluation of an event or situation.[13] No hemispheric asymmetries were observed in the central regions above the motor stripe. This asymmetry was compared to another facial expression involving five muscles, after an anonymous reviewer suggested whether this might be caused by the different number of activated muscles, two versus one. It was shown that the hemispheric asymmetry was not caused by the number of muscles involved, but by the emotionally positive mood association assumed from the earlier studies.

7.2.3.2 Analysis of the Experiment

This experiment shows the interaction of the internal as well as the external thought part, which was described in Chap. 6 (6.1.3.2, Thought Definition). The information expressed in the facial musculature by feelings, in this case more positive ones, is triggered by the internal elements, as is the case, for example, by seeing positive events. Conversely, it is possible that by consciously generating facial expressions associated with positive emotions, if done properly, similar activity patterns can be generated in the brain.

By consciously creating a certain physical posture, in particular the creation of a cheerful facial expression, an outwardly positive state was created for the test person. This effect was shown in the cortical activation and was similar to the

[13] Ekman and Davidson refer to different studies to which they contributed. These studies worked out different activation patterns of the hemisphere referring to watching movies or recalling or imaging emotional events. They further showed that a positive associated activation increased the patterns in the left anterior cortex region whereas a negative one increased the activity in the right anterior cortex (Ekman and Davidson 1993, p. 342f.).

expression. Because the mood was not measured directly, it is not possible to say whether and how the subjects perceived this change in brain activity. The fact is, however, that a change has taken place. This change corresponded to the expected expression of positive feelings. Furthermore, the other smile proved that the change in brain activity was not the result of the performance and thus the evaluation of the situation in general. A comparison with a more muscular facial expression confirmed that the number of muscles was not responsible for the different activity in the brain. Thus, the cause in the activity of the brain associated with positive associations is limited to the conscious induction of specific facial expressions.

As already mentioned, expressions triggered by emotions are regarded as part of the inner thought process. It should be noted that these can very well be of the most varied kind. Thus, certain expressions are universally valid, such as smiles and their positive aspects, and there are also many expressions of body language that have developed culturally (cf. Pease and Pease 2004, p. 10ff.). Here a distinction can be made between presumably genetically conditioned expressions as well as learned ones. Because the smile is regarded as a universal expression of positive emotions, it can also be assumed that cultural restrictions influence these emotions, but not in the same way as differently shaped gestures or even verbal communication.

This experiment shows that an emotional change can be created by changing body signals. This is all the more interesting because it is known from connectivity studies that this is not the case with so-called cognitive approaches, such as behavioural therapy. Thus, emotion can dominate rationality, but not vice versa (cf. Roth and Strüber 2014, p. 348). A change of emotions is therefore not possible through the understanding conveyed in the cognitive approach, but in a different way.

In Sect. 7.1.7 it was recognised that a change of the emotional colouring stimulates the creation of new things or even activates other areas. In connection with this finding, this experiment makes it clear that an activation strand influencing the emotional areas should be present starting from the physical signals. Thus, the areas involved in the processing of motor sensory input signals are likely to be upstream of the emotional ones in certain aspects.

By stimulating the correct body signals, switching states in the limbic system change. This order of interconnection can now also clarify how stimulation of the trigeminal nerve can lead to a change in depressive symptoms in PTSD patients (cf. Cook et al. 2016, abstract). In this case the stimulation leads to the activation of an emotional activity pattern, which was not connected with the negative emotional ones until now. Because it is regularly activated, it is associated with other related activity patterns. Because these stress patients had problems with a calm sleep, it can be assumed that the regular and calm stimulation of this nerve innervates via the path to the emotional centre that is also present in the smile. The patterns stimulated by this change the synaptic pathways through their regularity and can develop new paths or activate other patterns more easily. Thus, the emotionally stored associations are influenced by a change in both LTP and LTD to such an extent that this can have a positive effect up to a highly significant change in the depression symptoms.

The influence of the body on the emotional mood was also shown by other experiments in which the natural smile was induced by holding a pencil in the mouth, in which the lips were not allowed to touch the pencil or could only be fixed with the lips in the control arrangement. Related to their amusement value, a rating of cartoons in this case was substantially better for the smile, such as a variant other than a pout-like control version (cf. Strack et al.1988, passim). Another experiment showed a shift in the evaluation of prejudices related to ethnicity, in which faces presented with an artificial smile showed less distortion by skin colour in an implicit evaluation carried out via associations (cf. Ito et al. 2006, abstract).

With regard to the smile, it is confirmed that there is a significant distinction between a "right" and a "superimposed" one, which can be distinguished anatomically. This result shows that the actual emotion is expressed differently, namely, by involving the muscles of the orbicularis oculi. Such a distinction expresses that different associations can be associated with these facial expressions. In the discussion with the chunks it has already been explained that they are defined by so-called isa-pointers and thus more or less relate all experienced and associated as well as integrated elements with these in context. This reasoning has led to the concept of the "word cloud" created by the author, which is characterised by the fact that the different isa-pointers with the chunk are pronounced, fleeting, or more consistent depending on the degree of experience and importance. From this experimental arrangement it emerges that the emotions also have a corresponding sphere of influence in both directions. Therefore, these are also to be seen as a part of the isa-pointer of chunks, whereby they are more likely to lead to them than away from them.

This finding would theoretically mean that normally, without a change of mood taking place via the innervating path out, this would be formative for the activated chunks and their associations, thus also for the emotional colouring underlying an interpretation.

This finding is already known from experimental investigations. For example, in people who are hypnotised to certain moods, those experiences which are congruent with these moods are better reproduced. This state also affects the reproduction of word lists (cf. Bower 1981, abstract). These findings were also confirmed in reviews and showed that the material that was more congruent with the current mood from the point of view of affect could be more easily produced from memory. They also confirmed that the congruence and intensity of these affects had a positive effect on learning (cf. Singer and Salovey 1988, pp. 211, 245f.). This influence of mood goes so far that it affects the perception of facial expressions in mood situations induced by music. The subjects attuned by depressive music showed a tendency to shift neutral facial expressions towards fear in comparison to a control group, just as they showed more sadness and rejection in faces with unclear expressions (cf. Bouhuys et al. 1995, abstract). There are studies that confirm this influence, but they are limited to the recognition of faces and the language range for emotional words (cf. Chepenik et al. 2007, pp. 802, 809). This result would not question the derived functionality and its influence on the chunks but would limit them to a certain range;

it would also be consistent with regard to the specialisation of areas as well as the associated limitation of association and interaction between the various areas.

In sum, it can be stated that emotional processes have an effect on the processing of memory content. By these they have influence in the interaction on the negotiation of competitions, if they serve only on one side as amplifiers. Because in this experiment it was found that the emotional networks at least in their activity can be influenced by changes in body signals through a conscious change in posture, it can be assumed that there is a possibility of shifting this amplifier in negotiations and can thus lead to new insights and views or results. This possibility is conceivable in the direction of both positive and negative moods. Carson's views are thus confirmed in the possibility of creating something new in the "transform brainset," which is described as a consciously dissatisfied and slightly negative state of activity (cf. Carson 2010, pp. 208, 211), but limited only to the negative side.

Here, too, it seems to confirm what could be expected from the theoretical derivations. Emotions influence the thoughts and, by changing them, enable other activation patterns, which should be able to create something new.

7.2.4 Alcohol as a Context Influence

As early as 1969, interesting aspects of memory performance were found through findings from structured interviews with alcohol consumption. The survey of 64 alcoholics showed that in the course of so-called blackouts two different forms of memory loss seem to exist. The first of these can be traced back to the clear demarcation to the memory of other events of the evening and the sudden appearance as a "en bloc" loss and is probably caused by the physiological disorders of the brain, which was also expressed in the fact that it was no longer producible. Interesting, however, is the second type of memory loss: this occurred insidiously and was not actually registered by the alcoholics until they were informed of it, or until it was pointed out, or they were reminded and slowly created a reconstruction of this "forgotten" situations. As this reconstruction could be made easier by drinking again, it can be assumed that in this case a special kind of "state-dependent effect" seems to exist (cf. Goodwin et al. 1969, abstract). Because this effect can also be observed in everyday life, when alcoholics in a sober state no longer remember where they hid money or alcohol under the influence of alcohol, but are able to remember again by drinking, the influence of alcohol seems to trigger a special influence on memory (cf. Goodwin et al. 1969, p. 1359f.). This is also interesting insofar as for creative people alcohol seems to create an effect to stimulate creativity and incubation (cf. Carson 2010, p. 85f.).

7.2.4.1 Experiment Description[14]

This study involved 48 medical students who were divided into four groups for a two-part test procedure. Between the two tests, in which the same tests were performed, 24 h elapsed. The conditions tested were drunk (A) and sober (S). Together with the two parts of the procedure, the groups were thus able to take on the following characteristics: AA, AS, SS, and SA, with the first place always standing for the condition on the first day and the second for that on the second day.

Alcohol was administered by ingestion of 8 to 10 ounces (250–300 ml) of vodka (80%), which was mixed in a "soft drink" that was allocated according to body weight. For the sober groups, only the pure soft drink was administered in the same quantity. These drinks had to be drunk within an hour. The blood alcohol concentration was then recorded by measuring the air of the breath. The values were between 80 and 140 mg per 100 ml, with a median of 111 mg.[15] The test persons knew in advance that under certain circumstances alcohol would be administered to them, but this was the only information about the experiment they were given. The test series was carried out for all test persons in the same order over a period of 40 min.

An avoidance test was carried out as a test, in which the test persons had to suppress the lighting of one or more of four lamps. They were provided with a switch for each extremity, which served as a circuit breaker for one lamp each. An unpleasant tone was activated when an error occurred. The aim was to give 20 correct answers. The number of errors was used as the measured variable. The task was set the same on both days, whereby the arrangement of the switches was changed on the second day, that is, the switches led to different lamps than on the first day. This setup should express an interference effect in the error rate, which would lead to an increase in values compared to the first day.

As a further task, a (memory-)learning task was set. The test subjects had to learn four five-word sentences, one normal, one irregular, one anagram, and one pure word list. On the second day the test persons had to reproduce these sentences. After playback, the sentences had to be learned again. The measured variable was the number of errors in both sequence and completeness.

Subsequently, a word association test was carried out. Ten words with a low association value were presented, to which the test persons were to respond with the first word that came to their mind. On the second day, the stimuli were presented again and the test persons were asked to remember their answer from the previous day. In this case, the error number from day two was used as the measured variable.

[14] The experimental description as well as the results are from a study carried out by Goodwin and colleagues (Goodwin et al. 1969).

[15] The blood alcohol level was between 0.8 and 1.4 per milliliter. At a level of 0.8/ml and above, the first signs of drunkenness occur. Additionally, reaction time as well as perception already are influenced. At 1/ml usually speaking and orientation disorders arise and the ability to accept criticism decreases (cf. Kreilhuber 2014, Auflistung der Alkoholwirkung).

The next task was an image recognition test. The test persons were shown 20 pictures on the first day. On the following day, they should select these 20 from 40 pictures to recognise them. The pictures were divided into neutral pictures, which were models from order catalogues, and emotional pictures. The latter were selected from nude magazines.

The results showed that on the first day the sober groups were better in the learning task and the avoidance task than the alcohol groups, which can be attributed to the dampening effect of the alcohol.

By combining the two daily results with the group state and thus with the relationship of the change status, it was shown that for the change in performance in the two tests mentioned, as with the word association test, it was above all the change in state as an influencing factor on the change in performance that affected the change in performance.

In detail, those groups that had to change their condition showed fewer errors in the avoidance task than those that remained the same, meaning that they had less interference from the previous day, which in turn means that the performance of the previous day had less effect, that is, was less retrieved.

In the learning task as well as in the word association test, the AS group performed significantly worse than AA or SS did. The group SA deteriorated in its performance, but only from the value sober of the SS group on the first day to the value drunk of the group AA on the second day.

In the image recognition task, no correlations were found between the state changes in the neutral images. In the emotional images, there was a slight expression in the direction of state change dependence. However, this was not significant.

This experiment shows a difference between the state and the change of state, which suggests that the combination of information learned under the influence of alcohol seems to exist as a function of the state of retrieval. Furthermore, it turned out that soberly learned information is stored better than that not soberly learned. Finally, it became apparent that different contents might be influenced differently. The individual experiences seem to be more susceptible to this influence than those that already seem to have recourse to larger empirical values in the background.

7.2.4.2 Analysis of the Experiment

Why this experiment became part of the investigation has already been explained at the beginning of Sect. 7.2.4. On the one hand, the administration of alcohol seems to lead to effects in the storage of information, which can have a lasting effect up to the generation of context-similar frameworks. On the other hand, alcohol is sometimes seemingly seen by creative people as a means leading to or supporting creativity. At the same time, alcohol is an externally added substance that appears to cause changes in neuronal activity. Thus, analogous to others, it can be referred to under the term "neuroenhancer,"[16] which influence neurologically and thus, at least in their basic outlines, provide approaches for their mode of action in conjunction with the model. Alcohol is thus only one of many possibilities of stimulants that can interfere with and influence brain activities in a specific way.

This experiment focuses on the influence of the alcohol-induced state on memory. On the first day, between the drunken and the sober state, it becomes apparent that alcohol consistently leads to limitations in cognitive performance. For example, the median number of errors in the avoidance task is one third higher until the goal of the 20 correct answers to be achieved is reached. It is not clear from the data of the experiment whether this is noticeable in a slower learning performance at the beginning or in an increased error rate over the entire period. In this context, however, the task of memorising shows that the alcohol had greatly reduced performance. In summary, these two results confirm that the subjects' performance was damped by alcohol.

However, the experiments on the second day postpone these findings. The higher error susceptibility of the groups AA and SS to the state change groups in the avoidance task is associated with the interference of the previous day caused by the changed switch assignment. In this sense, the equality of state seems to lead to a similar effect as would be the case, for example, with the emotional context. Alcohol could thus, in a certain way, create a special kind of contextualisation. However, this is not reflected in the coloration as part of the isa-pointer (apart from the emotional states possibly shifted under the influence of alcohol), but should lead to a shift in the use of the memory traces caused by the influence of alcohol.

This derivation of the shift of the memory traces is strongly reflected in the results of the word association test:

- AA: 2.50 ± 1.57
- SS: 1.25 ± 2.55
- AS: 4.58 ± 2.13
- SA: 2.25 ± 1.43

These values are the error values of the associations with ten words from the previous day. It is obvious that in the sober state of ten associations, one in eight was not correct. In comparison, the associations generated in the alcoholised state with their retrieval in the alcoholised state led to an increase of the error values to double, now a quarter. It can therefore be concluded that under the influence of alcohol, associations produced under sober conditions are twice as bad as those produced

[16] Neuroenhancers are drugs or other substances used to increase the cognitive performance potential. These substances were derived from insights of neuro(biological) sciences as well as from medical sciences and used analogously in other areas (cf. Kennedy 2004). The use of these substances is discussed widely, including moral ethical aspects in an already performance-focused society as well as questions on physiological long-term effects and their impact in these frames in combination with stress/arousal research (cf. Shansky et al. 2013, abstract). Moreover, epigenetic consequences or societal aftermaths in the sense of class struggles or class discrimination are difficult to take into account. Despite these constraints, this knowledge is available and used in different circles of society (cf. Franke and Lieb 2010, p. 853f.).

under sober conditions. If one assumes, however, that the associations have not been noticed, but have only been created according to the task, and that the memory only shows whether this association carried out on the following day with a longer period of reflection would be able to evoke again the association created the day before, then one comes to the realisation of the reference of the associations with the strength of the memory traces. With regard to the isa-pointer already described by the chunks and their knowledge created by the experience or use of different strengths, a word association test is to be understood in such a way that it actually expresses the connection strength of the association. The first word association triggered by one stimulus is stronger than the other possible ones. Because dependence on the environment, that is, the context, is also an important factor, this is also taken into account. In this experiment, however, the distinction of the context from the same experimental framework is shortened from this equation except for the degree of alcoholisation. Therefore, it should be assumed that the restoration of associations should be relatively simple and thus successful, if in the reproduction phase on day two of the factor first association and thus the time pressure disappears, should be relatively simple and accordingly successful; this was also the case with the SS group. Group AA was only half as successful in comparison, but this could also be because of the memory shift by alcohol. However, the results of the SA and AS groups now become interesting from this point of view. The SA group achieved approximately the same results as the AA group. Considering that reproduction takes place without time pressure, alcohol also seems to interfere with performance in this case, but still leads to a quite respectable result. The median error rate is less than a quarter. It was therefore possible for the test persons to restore the associations of the previous day without the same time pressure. However, the AS group had a different situation and this was only able to reconstruct half of it. If one sees alcohol in this context as a means of changing the traces of memory, this result becomes clear. The first associations will be those of the test person walking on completely different traces of memory in an alcoholised state, who sometimes move far away from those in a sober state. On the following day, the sober state prevents the reestablishment of distant associations, because these memory traces are not strong enough to follow an activation and to be associated with it. Therefore, the associations produced in the sober state, both quantitatively and qualitatively, deviate correspondingly far from those mentioned the day before. This effect will not take place in the SA group in the same way, as in this group the associations delivered under sober state, which is also the everyday life and therefore can be regarded as normal, are strong. These associations can also be brought to the fore in an alcoholised state, although this would also produce the shift, with a little more effort and time, and thus be remembered or at least mentioned. But how can this shift in memory traces be explained by alcohol?

In its capacity as a substance dissolved in the bloodstream, alcohol not only has the ability to diffuse directly through the blood–brain barrier into the brain, but also has a special effect. Alcohol binds to $GABA_A$ receptors and thus reduces their activity (cf. Davies 2003, p. 263). GABA (γ-aminobutyric acid) is a transmitter substance used especially by the inhibitory synapses. Another indicator for the blockade of $GABA_A$ receptors by alcohol could be the reduction of $GABA_A$ receptor α-1 subunits, which occurs during long-term consumption of alcohol (cf. Davies 2003, p. 267). In that context, it is apparent that it is not only $GABA_A$ receptors in general, but by these different variants as well as subsystems, are each affected in a different way. This finding is confirmed by the variance of consumption quantities, which appears to be influenced by subsystems (cf. Davies 2003, p. 268) and also shows the effects of alcohol (cf. Pierruci-Lagha et al. 2005, pp. 1193, 1200f.). However, those partially genetic predispositions and differences have in common that they all influence the effects of alcohol on the $GABA_A$-receptors, that is, the inhibitory synapses; this is also an indication of the change in the traces of memory. If one thinks that chunks have a certain structure in their "word cloud," then this is imaginable by the competing form in relation to each other and in a context-dependent cascade. These chunks are structured among themselves by the interconnection structure in both an exciting and inhibitory way and intended for normal everyday use. By blocking certain neuronal networking structures, especially the inhibiting ones, the interconnection changes temporarily: this moves the negotiation patterns as well as the activation patterns. These are sometimes very different from those in the uninfluenced state, as inhibitory influences are eliminated or weakened. It thus becomes clear why, from this point of view, the AS group seems to see completely different results as associably understandable and why they can be traced back to the shift in the traces of memory. Alcohol per se does not affect the chunks as a context factor, but rather the shifting of memory traces through reduction of inhibitory synaptic activity. But how does it fit into this picture that the shift in the performance of the emotional pictures in the image recognition task has definitely had an effect on the change in the recognition performance?

The emotional colouring of a situation as well as the contents of memory in this context have an effect on the chunk development as a special part of contextualisation. The result of this experiment showed in the emotionally activating images, in contrast to the neutral ones, a tendency that is influenced by the change of state. Because alcohol, as a mood-changing substance, also affects the reward centres (cf. Davies 2003, p. 267), consumption produces a change in mood. As such, it is understandable that a change in mood results in a change in the performance of the reminder task. The fact that this result was not shown in the neutral images could serve as an indicator that the context information emotion had only a correspondingly low relevance when these neutral images were stored, which was not the case with the emotional images.

From the previous derivations it could be concluded that pathways altered by alcohol in the brain are used. One effect arising in this context is disinhibition, which could also be expressed in a change in self-control and thus in a change in hierarchical control. The determination of the reduction of inhibitory synaptic activity may have a role in why creative personalities experience positive effects in the incubation phase of creative work with alcohol. Because the shift of the memory traces as well as the reduction of the higher hierarchical control lead to the expectation of a state that produces alternative thoughts, as well as a better flow of these could be presumed, such an explanatory approach could be found.

However, by looking at this together with the previous findings, further thoughts arise. For example, it is interesting to note that five of eight US Nobel Prize winners in literature were alcoholics, as were many other creative people (cf. Carson 2010, p. 86). Because the memory of the associations found under the influence of alcohol is less likely to be retrieved in a sober state, but comes out more easily when this state is restored, it is tempting for a creative person seeking this state to go deeper and deeper into it. A tendency towards dependence derived from this would be understandable if it were not attributed to alcohol, but to its disinhibitory and thus creativity-increasing and new associative quality. This derivation is, however, only one part of various aspects which, in sum, could connect such an inclination with creative persons. Nevertheless, it should not be ignored that the group AA cannot only be associated with the shift of memory traces in the performance of word association, but also that the experiences taking place under the influence of alcohol are more strongly present in the drunken state than in the sober state. Part of the realisation of the avoidance task is that above all the equality of states was essentially responsible for the interference effect. This finding is interesting with regard to the alcoholic tendency, because it can be assumed that many of the memories created in the drunken state were not easily retrievable in the sober state, especially as this effect could probably lead to an increase in the dividing line over a longer period of time. Thus, working on efforts would be a problem for these persons if they were to work on them in different states that cannot be linked to each other.

The possible effect of alcohol on lateral disinhibition caused by the reduction of inhibitory synaptic activity has already been discussed. This aspect also concerns the area of attention. For these, it can be assumed that in this context a change in the focus because of the memory traces is to be expected and that this is subject to the reduction of the inhibitory performance, whereby a reduction in the focus and thus in the inhibitory control effect is presumably expected. However, this is only given from this experiment under the not clearly outlined approaches of memory. Because memorisation, as are all forms of learning, can be associated with conscious concentration on the subject to be learned, the poorer learning performance may be caused by a lower concentration or a poorer focus: the intention differs less from the incident than, above all, the depth of processing in the quality of retention (cf. Anderson 1996a, p. 190). However, this would have to be considered even more closely, although this aspect would also seem to be consistent with the expected result.

In this experiment, the aspect of using less frequently used memory content, which was also mentioned as a link for creating something new, is primarily generated by changing the memory traces used. This effect is probably all the more so in this case as it would be in other variants, because the inhibition of neurons (formations) is an important performance in the normal course of cognitive processes, which is omitted in alcoholisation. Thus, a substantial increase for this aspect is to be assumed by this change, which could be additionally strengthened by the mentioned emotional colouring as well as a shift in the level reference.

If neurogenesis, and thus the lasting effect of these effects, is not expressed in altered memory traces via LTP/LTD, this is reduced for the integration of something new. As alcohol has a negative effect on the formation of new neurons and their survival (cf. Shors 2009, p. 52), it can be assumed that this would not only reduce the ability to perform and integrate in the long term, but that these effects would also lead to health disorders and a reduction in plasticity (cf. Kempermann et al. 2010, p. 1). Because a long-term reduced neurogenesis expresses itself in a reduced hippocampal volume and the already mentioned reduced ability to integrate in other processes is associated with depressive episodes (cf. Roth and Strüber 2014, p. 262), it should come as no surprise that these images can be associated with alcoholics as well as with burnout, which is a special form of depression and is often associated with alcohol. De Bellis and colleagues, for example, have demonstrated these connections in an MRI study. In the course of the investigation, it was found that there was a significant reduction of the hippocampus on both the right and left side in subjects with abusive alcohol consumption (cf. De Bellis et al. 2000, abstract and discussion).

In summary, it can be said that although alcohol, similar to other substances, has an effect on the work of the brain and thus the potential for change and thus the creation of something new, it has an effect both in terms of a change in emotional coloration and in terms of a shift in the memory traces used, caused by a reduction in the release of inhibitory transmitters, and in terms of a change in the level reference and a decrease in the inhibitory effect of attention. However, changes, including dependencies, are as likely as long-term damage, which ultimately even worsens the creation of something new through integration prevention or reduction in the new experience. The basic realisation, however, is that the approaches found in the model of thinking could also be shown here.

7.2.5 Meditation as a Means of Shifting Thoughts

Meditation is a special technique in which the focus of attention is trained. This very introspective mental activity differs from normal everyday life. The effectiveness of meditation has been confirmed by many studies, but the significance as well as the question of why is unclear. This lack of clarity is also attributed to the lack of standardised and comparable designs. Nevertheless, it can be seen that better performance is achieved in concentration tasks, that activating situations are reacted to more calmly, and that stress reactions decrease faster. All in all, meditation shows that it has an effect on the function of the central nervous system (cf. Cahn and Polich 2006, p. 200ff.).

A review from 2014, in which Marchand considered studies on the subject of mindfulness, came to similar conclusions. Mindfulness is an aspect that is also practiced in various teachings and techniques of meditation. It is an introspective attention, consciously concentrated on different areas, which should enable a fuller experience. The aim is to achieve a dispassionate state of consciousness that is focused at the moment, which can include emotions, perceptions, and sensations as well as thoughts (cf. Marchand 2014, p. 471). These states have shown increases in focusing in various studies, which can be seen in the areas of the right frontal cortex as well as in the dorsolateral prefrontal cortex and the anterior cingulate cortex (cf. Marchand 2014, p. 474), but the effects of mental wandering are also shown. This wandering uses the "default mode network," which is associated with automated thought processes that are difficult to control. Mindfulness should lead to more conscious perceptions in this area, but also less irrational negative interpreted feelings, which can reduce symptoms of depression, anxiety, or stress (cf. Marchand 2014, p. 475). Thus, it is shown that a change in cognitive processes can be produced by meditation and mindfulness, as well as a limiting assessment should be reduced. Consequently, it is to be expected that this could also be used to increase creativity. To look more closely at this, this chapter will examine an experiment that generates a change in thoughts through feedback to a creative process, which can then be presented as an example of an experimentally controlled eureka moment.

7.2.5.1 Experiment Description[17]

Two groups were investigated for this experiment. The first one was developed with the "integrated body-mind training" (IBMT) from Chinese medicine, which reaches meditative states through a special intervention of physical and mental activities. For the second group, relaxation training (RT) was used as an intervention method, in which the tested persons began to feel more warmth and a feeling of heaviness through the relaxation of muscles, as a result of which they relaxed more and more.

For the experiment, 32 right-handed students without experience in meditation or relaxation exercises were randomly divided into two groups. The groups were each given 30 min of training per day for 10 consecutive days, so that a total exercise duration of 5 h per testee was achieved.

The experiment used elements of the Chinese "Remote Associates Test,"[18] which, among other uses, can be used for creativity performance. Three problem words were shown to the test persons. They should find a single other word, with which together with each of the problem words a known Chinese phrase should be formable. The subjects had up to 19 s to do this. If they found a solution before, they should push a button. Then followed the sequence of the standard answer, which followed either after pressing a button or 19 s display of the problem words. The standard responses should be assessed as yes or no after a 3-second presentation, with the assessment aimed at classifying the standard response as better fitting than the own variant found to solve the three-word association. Because those persons

[17] The experimental description as well as the results are from a study carried out by Ding and colleagues (cf. Ding et al. 2015).

[18] This is a modified version of the "Remote Associate Test" (RAT), which Mednick developed and tested in the 1960s. He found high correlations between creativity and performance in the test (cf. Mednick 1962; Mednick and Mednick 1967).

who had a different answer but classified it as more coherent, as well as those who did not find an answer but had now received a coherent answer, these circles of persons who answered "yes" received an "insight." This test sequence was recorded using fMRI. Because meditation uses different areas than relaxation techniques, it was expected that this would be reflected in the activity of the corresponding brain areas during these processes. Therefore, the RT is used as a control group of the IBMT experimental group and should make it possible to visualise the activities using fMRI. The fMRI values were processed by various standardised methods into a smoothed statistical data set with which the activation strengths over time as well as their changes in retrospect could be evaluated via individual voxels. This review is relevant to this experiment because it should express insight as a change in the assessment of the previous process in the neuronal activity pattern. In addition, other methods were used for evaluation, so that comparisons between test persons were also meaningful.

The process was divided into three phases. The first phase, pre-session, was used for the subjects to carry out the experiment to obtain initial reference values. This stage was followed by the 10-day "training session." The groups performed the exercises in the respective training variant with the help of a coach or a CD guide. The last section was the "post-session" phase. In the latter, the experiment in the scanner was carried out again, where the degree of difficulty was the same, but other examples were used to reduce the occurrence of similarities.

The results of this experiment showed in the "pre-session" no differences in problem-solving ability between the groups. These values in the "post-session" were remarkably different, so that it can be said that the IBMT had induced significantly better creative problem-solving performance than the RT in terms of short-term effects.

With regard to the imaging procedures, in the "pre-session" there are no significant differences in activation between the two groups. In the "post-session," phase differences were observed in the right cinguli gyrus, the insula, the putamen, the right inferior frontal gyrus, and on both sides in the medial frontal gyrus as well as in the inferior parietal lobe and superior temporal gyrus. In these areas, IBMT was more active than RT. The RT group even had a decline in activity in these areas compared to "pre-session." In addition, the IBMT was more involved in restructuring the problem than RT through this 5-hour training.

The gyrus cinguli is associated with cognitive control and the ability to change the ideas for insight. The areas of the medial and right inferior frontal gyri correlate with a positive creative performance. Together with the inferior parietal lobe, they are associated with a renewed problem representation. All in all, this shows that the IBMT is in a better position to suppress the original idea and thus to better approve and integrate the standard solution. The areas of the insula and the superior temporal gyrus, together with parts of the inferior temporal lobe, are associated with increased self-awareness and introspection, as well as control of emotions, which in this case is associated with their role in error finding, problem understanding, and general attention control. The increased activity in putamen is associated with the release of dopamine, which is caused by the change from a dominant but false response to a correct one and therefore an "*Aha!* moment" is generated. This response is associated both with the recognizing change and with the reward centers often involved in meditations, and thus their easier activation.

The reduction of brain activity in RT may be caused by its need to produce conscious relaxation during relaxation exercises, for which relatively much effort is put into maintaining relaxation. Under certain circumstances, this could even be reflected in the reduction of the resources necessary for the insight task and thus in the low level of activity in the areas necessary for it.

7.2.5.2 Excursus Integrative Body–Mind Training

To take a closer look at this experiment, here is a short excursus to explain the integrative body–mind training. This training includes physical relaxation, mental presentation, and mindfulness training, which is usually combined with specially selected background music. Through introductory instructions, a starting mental state suitable for the training is achieved without the control of thoughts; it becomes much more a state of relaxed "alertness" that can be generated. Through this an increased awareness about body, breathing, and external instructions is given (cf. Tang 2011, p. 383).

The IBMT already leads to a change in the circuit structure after short training units and a short total exercise time (cf. Tang et al. 2010, p. 15649). Among other things, this happens conspicuously in the anterior cingulate cortex (cf. Tang et al. 2010, p. 15650f.).

7.2.5.3 Analysis of the Experiment

The results of the experiment suggest that meditation can show an increase in creativity. Although it was already stated at the beginning that this experiment seems to induce a eureka moment through feedback loops, as was also shown in a cross-reference given by Ding in a study by Siegler in 2000 by the sudden change from a deadlocked view to the solution of the problem, in which young students were examined in solving a mathematical task and described the solution approach. In the course of the study, it became apparent that some pupils unconsciously used an arithmetic abbreviation and that it was only possible to articulate its presence in consciousness after a few rounds (cf. Siegler 2000, pp. 79, 82). Nevertheless, in this meditation experiment it is initiated from the outside and not from the inside, so the path to the *Aha!* experience does not come from the inside, but is achieved through the standard solution.

With regard to the previous findings of this work that both postures in both directions can change brain activity and that a dream is a certain recapitulation of experienced events, triggered in a self-similar way in reverse order of storage, it can also be assumed here that through the similarity of the course of events the results are likely to be similar in a certain way to the effect in Siegler. If one also includes the fact that a significant increase in creativity can be associated with the IBMT, it can also be assumed that this effect would be present regardless of the type of impetus. In this second case it would only be important to distinguish that the retrograde Aha! moment would not be associated with the mentioned areas. However, this should not restrict the results, as already deduced from other aspects of this work. An additional aspect, which in this experiment speaks for the experience of the Aha! moment and thus for an idea, is the subjective evaluation of the subject in relation to the standard solution, if this deviated from his own. The test person assessed whether this standard solution appeared to be more valid and thus reevaluated the condition. Such a revaluation would have to activate similar processes in a certain way as the independent search despite the comparison with the own term. It can thus be successfully stated that the revaluation results in processes similar to those of the own search, so that similar areas are activated. In addition, by attributing the more valid evaluation of another term, these lead to an insight that finalises the creative process initiated from the outside in a state close to the Aha! experience, which, compared to the previous state, has produced something new. Thus, it can be assumed that this experiment is a conceivable kind of creativity in the sense of this work, which, combined with a specific meditation technique, leads to better results than in an alternative relaxation technique.

The state of relaxed alertness achieved by the IBMT leads to an improved perception both of the body and of thoughts or breathing. This shift in attention leads to other neuronal activities, bringing sensory activity patterns from the unnoticed area into focus, whereby they are likely to experience an increase in activation strength from lateral inhibition, that is, enhancing contrast. A similar phenomenon has already been described in Sect. 7.1.2. "Chunks thus represent an integration of sensory information in different depths and represent it in specialized areas." This finding shows in this case of the experiment that a shift of attention to otherwise unnoticed areas of both the body and its actions leads to the activation of stored but rarely used chunks. As these are defined by other "clouds of meaning" and associations, new traces of memory can be shaped and used from these.

Through repeated practice and strengthening of these processes, the memory traces triggered by these techniques change and become easier to activate. Sustainable meditation trainings thus change both the synaptic interconnection efficiency through LTP. Furthermore, this type of exercise reduces stress reactions such as heart rate and skin conductance (cf. Tang et al. 2009, p. 8866). This reduction in stress reactions leads to an increase in both neurogenesis and better processing during sleep, resulting in better storage and integration (cf. Roth and Strüber 2014, pp. 359 in conjunction with Spitzer 2009, pp. 121f., 125f.).

Thus, starting from small activated chunks, varying structures are newly strengthened by strengthening the memory traces emanating from them. This result was already confirmed by Tang and colleagues in 2010, as stated in Sect. 7.2.5.2. These changed cognitive activity patterns should also change the usual production rules for their use. Because the production rules are also in basic activation from the dependency existing in the respective topicality reference, this will lead to different framework conditions than would be the case in the accustomed environment.

In both the cases of chunks and in production rules, the changes mean that previously less conscious or used areas can be brought to the fore by shifting attention. These areas, which are actually sensory, change creativity to the extent that in other convergence zones they lead to associations that are unfamiliar. This is already expressed after 10 days in the significant increase of creativity performance in the Chinese RAT, which, according to its creator, is supposed to measure exactly this, namely, creativity.

Another effect created by the IBMT is, through the relative lack of evaluation, a reduction of emotional influences in hierarchical control. As has already become apparent from the areas of anxiety and depression, it is hardly possible to influence the amygdala purely through cognitive control because of the lack of innervation (cf. Roth and Strüber 2014, p. 348). However, because this type of training showed that stress values as well as symptoms of depression or anxiety can be sustainably improved (cf. Marchand 2014; Grossman et al. 2004; Majumdar et al. 2002), the influence could also occur in this case via the body. Similar to the experiment of the smile, a special posture together with the harmoniously calming associations would suggest that the signals coming from the body, analogous to the trigeminal, could produce this relaxing effect.

In addition, one goal of meditation is to create a state of nonevaluation, which would create a change in the hierarchy of control. This change would be supported by the fact that the IBMT showed an increased activity in the area of the cinguli gyrus compared to the RT, which, as the authors themselves had already stated, together with the medial and right inferior frontal gyri and the inferior parietal lobe, would be related to the re-representation of the problem and the associated displacement of the solution. In addition, the insula and the superior temporal gyrus are part of the areas associated with self-perception, introspection, and emotion control. In these areas, too, there was increased activity, which was shown to be essential in the importance of finding errors and understanding the problem, that is, in the acceptance of the possibility of an alternative solution and thus the acceptance of such an alternative solution. Because the imaging and thus the evaluation of these areas only took place when the standard response was considered to be better than one's own, this activation is specific for the adaptation in creativity. It can therefore be assumed that these activations are specific to the view. However, it is not known how these have a quantitative effect or to what extent the IBMT has resulted in a shift in the values of the answers in comparison with the RT. Only the reference to the significantly increased performance in the IBMT in this creativity test can be confirmed. Nevertheless, with regard to the change in hierarchical control, it is assumed that a change in openness can take place through the IBMT, as the areas associated with it from other studies were more strongly activated vis-à-vis the RT.

Mindfulness is the state of letting go of thoughts and allowing what comes. At the same time, a passive role of attention is assumed, which no longer ensures that the focus is on one area and thus an increasing contrast takes place between the various neuronal activities in the negotiating position, but rather takes on the partly equivalent, thus presumably oscillating, simply more equivalent activities coming from these negotiating positions. Thus, a flow between thoughts and states can occur, and this can be on a very high level of "unevaluation" and thus equality as a kind of emptiness, if these can no longer stand in any conflict for attention. Under this aspect, the change in hierarchical control can probably be associated with the fact that the conscious defocusing of attention reduces its effectiveness as a contrast enhancer or mediator. The attention could therefore have a further role than that of the recapitulation mentioned in Sect. 7.1.7. If attention is characterised by absence or strong reduction, this would support the reduction of hierarchical control. A closer look at various reports by creative personalities makes this easier to understand. It is often said in conditions of twilight or tiredness that sudden inspirations have taken place and thus laid the foundation for great progress (cf. Andreasen 2005, pp. 40, 43, 45).

In this context, the increased activity in putamen associated with the release of dopamine, which stands for the actual Aha! effect, may be interesting. The function of dopamine corresponds to motivation (cf. Roth and Strüber 2014, p. 101). It is also a signal substance for an unexpected reward stimulus, in both positive and negative respects (cf. Roth and Strüber 2014, p. 97). Thus, the activity in this experiment is just as understandable as it is in the generally increased involvement in meditation (cf. Ding et al. 2015, p. 47). In the experiment, the acceptance of the new result produced a surprising Aha! effect. Because it is accepted, this is accepted as positive and thus rewarding. In meditation exercises, on the other hand, various new experiences are experienced which result from meditative practice alone. It is understandable and coherent that this innovation would lead to an unexpected experience that would be accepted as a reward in a calm, accepting, and generally positive environment.

Considering the just mentioned peculiarity of the IBMT, the change of the level reference probably also has a part here, by paying attention to areas of the body that otherwise receive little attention. This point has already been discussed. However, the shifting of the reference level has a role in which fineness, that is, detailed movement or body excitation, is mentioned. In this case, the atypical observation of small and minute changes is something new. Not only to take care that you inhale and exhale, but also that you try to feel the path of the breathed air through the nostrils, further exchanging the air within the inside of the head, past the throat into the neck. One tries to feel how the air can flow past the larynx and slowly penetrate into the chest, from which it stretches from the belly over the chest and then the flanks. If one tries to follow this listing of the way of breathing air in contrast to the abstract statement—one breathes in—then it might become obvious that the level has shifted from the abstract to the detailed. This concept is similar to the analysis of sentences whose meaning is sought, or the syllable analysis of these sentences to explore how they produce a word in which context with linguistic changes. The example of breathing can be extended and applied to many other areas for such a meditation technique, as it is also used in practice. It should be noted that these targeted observations are only partly stimulated to achieve a relaxed state, but should simply be accepted in the phase of relaxed alertness.

Although the aspect of expertise could play a role in this context, especially if one can speak of many years of intensive experience and practice on the part of the persons involved, this experiment is not meaningful for this purpose, because only a short training was deliberately used as a distinguishing feature to keep any other influences as low as possible.

Thus, it can be summarised that the areas addressed in this experiment, which have led to an increase in creativity, can probably be found in the better acceptance and reduction of hierarchical control. This stage is achieved by a defocused attention, which is unambiguous in this experiment. In addition, the type of training should activate unfamiliar chunks, memory traces, and production rules through the new attention focuses, which are otherwise rarely used. An important aspect here is the shift of the reference level, which changes from the macro to the micro level. These areas are likely to be subject to an LTP for the duration of the training, which makes the changes sustainable and at the same time more likely to be activated. Finally, one can once again mention the defocused attention, which, even as relaxed alertness, should lead to a change in hierarchical control, above all through a nonevaluative state, and thus to a better ability to integrate, but also to a more open attitude towards something new.

But why are these effects not achieved to the same extent in RT? The RT were carried out in such a way that the subjects devoted themselves to progressive muscle relaxation after Jacobsen. This technique, taken from the manual by Bernstein and Borkovec, uses the principle of tension and relaxation by progressively stretching individual larger muscle groups, such as the upper arms, from the head to the upper body and extremities after a specific ritual, starting with feeling, and then consciously relaxing. This stage is followed by "sensation focus," which in sum leads to relaxation of body and mind. The experiment focused on the feeling of warmth and heaviness.

An essential difference of the trainings is the focus in the attention. As the RT expresses itself on the respective areas of the body in a tension–relaxation focus, it is targeted and no open states are created. This thinking in controlling mechanisms is also assumed by the authors of the experiment to interfere with the training effects, which is why they regard RT as a suitable control group (cf. Ding et al. 2015, p. 43). The feeling as well as the feeling before or after the tension is similar to the state in the IBMT when looking at different areas of the body in detail. However, it differs in that the observations controlled by conscious processes lie in the areas to be tensioned. This step does not create a purely passive perception, but an active-passive perception that focuses attention on a special, closely framed area. The differentiation of the groups by this effect, suspected by Ding and colleagues, could be expressed in the nonincreased activation in the areas, which are connected with openness and re-representation of the problem.

A further aspect with regard to the change in the level reference could be that in the course of RT the focus is on large and generally heavily used muscle groups, which are also easy to control. For example, when tensioning the thighs, the focus is easier on the biceps femoris as well as quadriceps femoris, whereas the areas of sartorius muscle, semitendinosus muscle, and the remaining six muscles are hardly considered. To what extent the higher resolution of the observation and thus the detail accuracy through the shift of the focus can contribute to this is to be interpreted from the model of thinking in such a way that through the instruction to concentrate on the large and frequently used muscle groups, leads to the activation of the chunks corresponding to their common way of use. This approach would also suggest that the chunks, memory traces, and production rules would change their use analogously to the IBMT, but that it would be less pronounced. Whether this analysis can explain why there is also an improvement in the accuracy of the "Remote Associates Test" cannot be explained in detail here, because the author cannot access the data. However, he would assume that the improvement would be caused by this circumstance and not only a possible change in the sense of the statistical fluctuation band.

This reference to the summary of the influences likely to change in the IBMT through approaches from the model of thinking with relaxation training should thus confirm and sharpen the points and effects mentioned. Therefore, in this case, too, the author considers the parts of the model of thought to be probably coherent by closer examination.

7.2.6 Hermeneutic Circle as Creativity Technique

An essential assumption of this work is the emergence of memory contents, however they may be constructed, which emerge from the distorted sensory entries and their increasing association and reference among each other. Because in this context they represent a phenomenon triggered by the environment, but this cannot be seen without its own influence on it, hermeneutics with the distinction of a prior understanding preceding understanding (cf. Mogge-Grotjahn 2011, p. 22) is essential. The prior understanding influences not only the understanding of an interaction with the environment, but also the further understanding of it through this interaction. In this context, the hermeneutic circle shows that prior understanding and understanding are two interrelated parts that are interdependent. The model that emerged from this work now offers a preliminary understanding that enables further consideration of this hermeneutic circle.

7.2.6.1 Description of the Hermeneutic Circle

To be able to analyse the hermeneutic circle from the point of view of the model of thought, a brief description follows in this chapter, and thus an attempt is made to explain it so that the essential aspects and processes are known.

As already mentioned at the beginning of the superordinate chapter, it is part of the hermeneutic circle that there is a prior understanding. This is part of the person who wants to deal with a subject, be it a book, a film, or a discipline of science. Already the fact that this person sees this topic as relevant to be engaged with shows that this person must have a prior knowledge or understanding of this topic. It would, for example, have been impossible for Christopher Columbus to know the significance of interacting with indigenous peoples on the American continent before sailing west. From his previous understanding, however, it was clear that a journey towards the west, since the assumption of the spherical shape of the earth was present, would have to lead to Asia. Asia corresponded to the knowledge of that time as the easternmost edge of the mainland. It was clear from these understandings that the discovery of mainland land at the end of his voyage led to the assumption that he was right and had now created a sea route to Asia. It is obvious that, from today's perspective, these two constructs would lead to other interpretations of such a crossing because of the existence of the American continent with an extended prior understanding. Nevertheless, today we still speak of the term "Indians," which comes from that time, when the indigenous peoples of North America are meant. Here a term was still preserved in its being generated by constructs against better knowledge. Prior understanding thus influences the process of recognition by limiting it to the circumstances possible within the framework of prior understanding.

The hermeneutic circle expresses that this influence of prior understanding in relation to the new, be it perceptions, insights or anything else, leads to the emergence of an understanding which results from the prior understanding and the new in an integration of the two. This does not have to be done in equal parts, nor does it have to be logical. A further and at the same time very important aspect is that this understanding becomes a prior understanding of the following experiences. This results in a circularity, which leads to a constant progress of a transformation of the understanding that emerges from prior understanding and the new into a prior understanding.

The reflection of this process[19] leads to the realisation of the underlying assumptions as well as their relativity becoming apparent. This understanding is part of the scientific theoretical debate on the question of the legitimation of knowledge per se as well as its achievement. In this analysis no reference is made to this, but an attempt is made to understand how this process works through the model of thinking.

7.2.6.2 Reference of the Hermeneutic Circle to the Model of Thought

The hermeneutic circle presupposes the existence of prior understanding. In the model of thinking, this is thought through both the chunks and the production rules: these include the contents of memory in their entirety, both as information in itself and in relation to one another. Thus, it can be assumed that these are consistent with the previous understanding in the sense of the hermeneutic circle. If, for example, specific chunks in the respective topicality reference are now determined as by this and are more likely to occur than others, this does not express a limitation of prior understanding in general, but in the specific. No protest can be lodged against this restriction in the hermeneutic circle, because it speaks of prior understanding in

[19] Because this process of reflection creates something new, the author ascribes this approach as similar to creative techniques.

general. Thus, the restriction of the prior understanding would only be to be understood in the sense that it should be seen in relation to the respective person, but not to the different contents existing in the person, which would otherwise always be thought of as comprehensive and occurring at the same time. Thus, the topicality reference of the memory contents does not restrict the validity of the hermeneutic circle.

Because the activity structure of the chunks comprises a spreading of the memory traces to the areas associated with them, not individual and narrowly framed areas are thought of in this context, but a complex of themes related to this topic. As the traces of memory are differently pronounced in them, it explains how these networked areas can be influenced to different degrees by a new topic.

If one now consistently thinks of the hermeneutic circle in the experience of a completely new one, then it follows that the integration of this hitherto completely unknown must also lead to a reference to a prior understanding. However, this prior understanding is not appropriate or valid with this really new one. It is actually by chance one that is related, which does not seem meaningful or realistic. Analogy can offer the approach of an explanation here. Analogies in this context are similarity relationships, which occur from the person resolving this conflict by reference to a new one with familiar old structures (cf. Rieken 2010, p. 193). Thus, similarity patterns are created to form a coherent part of the prior understanding as a connecting factor. This process would also be understandable from the model of thinking if it were thought from this that the activation of chunks expresses itself in their "word cloud" or "cloud of meaning." In this case the chunk, symbolically conceived as a cloud, would be activated by the highest topicality reference, because all others have even less reference to this new one. From this chunk, the corresponding isa-pointers, which have the most similar characteristics to the new, could be seen as the different densities of the cloud, which now, because of the hitherto unknown activation, move together in sometimes new tracks: this creates an analogy between the new and the familiar.

As this process sometimes does not happen immediately when something is completely new, various stimuli and supports are necessarily part of this process so that the threshold of consciousness can be crossed. Openness to new solutions is an important aspect here. If a topic is considered impossible by the person because of its nonintegration ability, the necessary time and activation would hardly take place. If, however, the openness is given, the observation of a fireplace can lead, as with Frederich Kekulé, to the penetration of an analogy that leads to a conclusive understanding of a problem[20] that he had long cherished and considered to be important.

Irrespective of the degree of novelty and the type of reference to the previous understanding, the new establishes a link with the existing. This connection will then, through integration, result in changes in the characteristics of traces of

[20] Frederich Kekulé found the answer to the question of the chemical structure of carbohydrate compounds, especially for benzol (C_6H_6), after he visualised it as a ring. The inspiration took place while he was napping in front of an ingle, which reminded him in his doze of twisting and winding snakes (cf. Andreasen 2005, p. 43f.).

memory. This leads to changes in the clouds of meaning, that is, also in the chunks themselves, which can occur both in the linking efficiency of synapses and in the integration of newly formed neurons. The change brought about by this adaptation creates a new network structure that corresponds to the combination of prior understanding and the new. Thus, prior understanding and the new were combined to form a new understanding, which again forms a basis for the following experiences as prior understanding. This prior understanding is expressed in a changed pattern of activity, which should find expression in the shift, involving the vectorial understanding of a thought, in treating the same situation from the first time. This shift and change confirm the mode of action of the hermeneutic circle through the model of thought, although neither the depth of the critique of cognitive ability nor an epistemology based on constructivism is addressed.

7.2.7 Chance as a Moment of Creativity

In the last paragraph of Sect. 7.1.6, Neuroanatomical approaches, it was indicated that chance could also be a component in the emergence of something new. This approach is further explained and exemplified by the author in this chapter.

In the mentioned chapter, an experiment by Guttmann showed that the physically identical stimulation of the human acoustic sensorium led to physically different stimulation provocations, which in this way showed the inadequacy of the environment in the sensory stimulation activities. One of the possibilities co-responsible for this by the author was seen in a kind of fatigue, which "could be expressed by the repetitive monotonous same external excitation in a delayed internal activation." The fundamental derivation of the possibility of this displacement and of any influencing part of it, the ion equilibrium, has already been explained in detail. Starting from the fundamental possibility of the collapse of the resting membrane potential no matter what the cause, whether this be the possibility mentioned earlier above or a pure coincidence or aspects of deprivation in the glucose[21] range, this chapter now considers a thought experiment.

To carry out this thought experiment, the sequence of triggering an action potential in a neuron must first be remembered. Basically, a neuron is charged with about −70 to −90 mV[22] by maintaining the so-called resting membrane potential. This is

[21] A study with mice showed that nutrition with high fat content changes the performance range of the brain, because the permeability of the blood–brain barrier adopts consequently to a nutrition-related change in the blood, so that less sugar is delivered into the brain, causing a loss of performance. The structural change reacting to a longer lasting fat-rich nourishment leads to an increase of the "vascular endothelial growth factor," which sustainably influences the absorbance quality of glucose and furthermore is associated with damage and diseases such as Alzheimer's (cf. Jais et al. 2016, abstract and highlights).

[22] Silbernagl and Despopoulos write in the same book (p. 44) of general cell types, nerves as well as muscles, about a resting membrane potential of −50 to −100 mV, in figure A1 (p. 47), or about −70 to −90 mV. As other sources describe the resting membrane potential within a range of −65

controlled by an Na^+–K^+–ATPase (i.e., an ion pump; cf. Silbernagl and Despopoulos 2003, p. 26) that is active and passively maintained with minimal unequal distributions (cf. Silbernagl and Despopoulos 2003, p. 44). By shifting this resting value in the direction of less negative values, that is, a predepolarisation taking place, the membrane potential approaches the threshold potential. Once this is reached, an action potential is generated that is characterised by the additional activation of further ion channels, which results in a rapid increase in the positively charged ions and has an effect on the associated collapse of the membrane potential. The inflow of the ions leads to an overshoot with peak values of +20 to +30 mV. During the development of this depolarisation, opposing mechanisms already begin to take effect, which lead to the restoration of the resting membrane potential in the shortest possible time. In this phase of repolarisation, the neuron is in a refractory phase. The refractory phase is a phase that is divided into an absolute refractory phase, in which the neuron is not excitable at all, and a relative refractory phase. In the relative refractory phase, the neuron is more difficult to depolarise than in the initial state. In addition, the triggering of an action potential in this phase leads to a lower level and steepness of increase of the action potential. The repolarisation cycle is completed by a short phase of post-hyperpolarisation in which negativity values higher than 90 mV are achieved. One reason for the refractory phase is the impermeability of the ion channels at certain voltage values. For example, for the Na^+ channel the optimum value for a maximum inflow is −100 mV, whereas at −60 mV a flow rate of only approximately 60% can be achieved, and from −50 mV this no longer can be activated (cf. Silbernagl and Despopoulos 2003, p. 46).

Another important piece of information for this experiment is that the Na–K pump is dependent on the energy supplier ATP to maintain its equilibrium as a transporter. A lack of energy or an inhibition of the Na–K pump leads to the collapse of the membrane potential (cf. Silbernagl and Despopoulos 2003, p. 44). The transportability of the Na^+–K^+-ATPase is relatively slow compared to fast ion flows. The value here is approximately 1:1000, whereby a shift in the ion concentration increases the pump rate of the ATPase (cf. Silbernagl and Despopoulos 2003, p. 26). In this context, it should also be mentioned that a sustained disruption of the Na^+–K^+-ATPase can lead to a lasting disruption of a cell, so that it can be brought to cell death (cf. Yu et al. 2003, p. 2099).

Action potentials last about 1 ms and can be triggered up to 1000 times per second in special nerve cells. Only one hundred thousandth of the intracellular ions flow per release (cf. Yu et al. 2003, p. 46).

The triggering of an action potential depends on several spatially or temporally summed, incoming and exciting potentials. This process must take place within 50 ms, because the predepolarisation has not yet decayed in this period. As a single incoming excitation potential can be shifted by a maximum of 20 mV, but the threshold potential is around −50 mV; the relationship between spatial and temporal

to −90 mV, the aforementioned value is used, particularly because for the thought experiment this detail is of lesser importance.

excitation is clear (cf. Yu et al. 2003, p. 52). This maximum value depends on how efficient is the connection between pre- and postsynaptic neuron.

In principle, it can therefore be assumed that a neuron, when depolarised, leads to the restoration of the resting membrane potential by blocking the ion channels with respect to the influx and increasing activity of the ATPases in the opposite direction, after a brief overshoot by the interaction of these same elements. At high excitation frequencies, the processes necessary for recovery, such as Na^{+}–K^{+}-ATPase as well as the processes required in the neuron for the restoration of the transmitter substances required for transport, are continuously carried out. ATP, which can be produced by a combination of water and glucose in the mitochondria, is required above all for these processes (cf. Yu et al. 2003, p. 12). As neurons possess very many mitochondria (cf. Kahle and Frotscher 2005, p. 22), it can be assumed that these cells require a great deal of energy. In the case of a high-frequency and monotonous or very self-similar stimulation, it can be assumed that temporary local bottlenecks could occur after a certain duration. Whether the release of the ions would be so high that their concentration would lead to a restriction, or whether it would be caused by a reduced efficiency of the ATPase, remains to be seen. Through intensive use it would be conceivable that the restoration of the ion balance could be disturbed in the short term. This disruption in turn could lead to a local collapse of the membrane potential, which would again trigger an action potential. Or, an incoming signal could be limited by the extended refractory phase and have no effect. In both cases an asynchrony would be generated. This asynchronicity can be directed both forward and backward, that is, it can be triggering or blocking.

An asynchronous activation has the effect of a temporal shift, which can extend to a few milliseconds, depending on how long it takes to restore one's balance. It can therefore be assumed that the local collapse of the membrane potential of a neuron by limiting the ATPase performance is conceivable. On the basis of this, it can also be assumed that such a neuron, which would be randomly predepolarised or would even become active on its own, can have an action potential that propagates along its own axon to one or more other neurons.

In Chap. 3 (3.2.2.6, Expertise) it was stated that this is characterised by the fact that the networks involved can be represented more densely and better in the brain, and have more and more comprehensive content and relationships with one another. Expertise thus expresses a particularly well networked and thus efficient neuronal structuring. In connection with long-term potentiation, expertise is also a topic that is subject to frequent use or a corresponding attribution of importance. In this second aspect it is expressed that a subliminal, unconscious pre-activation can express itself. This effect can often be experienced when, deep in a conversation, you hear a triggering word at another table, whereupon your attention suddenly shifts to the conversation in which that word was spoken.[23] With this effect, a special assessment

[23] This as cocktail-party effect known phenomenon usually is experienced with hearing one's own name (cf. Gerrig and Zimbardo 2007, p. 143 and 164), but also can occur with other topics.

should trigger a stimulus cascade from the incoming stimuli that results in a shift of attention. Part of this cascade could be that topics that occupy a person are kept up to date in the pre-conscious and could thus be more easily raised to a level above the threshold of consciousness. A similar effect can be observed with the incubation effect when solving problems. This effect is produced by turning away from a problem in which one is stuck, which leads to the fact that one is more likely to find a solution when the problem is focused again (cf. Anderson 1996a, p. 264). Anderson sees the reason for this in the fact that the memory traces "relax" and thus become more widely available again, but also the effects over an unconscious processing are still conceivable thereby. In any case, however, the excitability of neuronal networks is facilitated in both expertise and current topics.

If now the conceivable random triggering of an action potential from a neuron takes place, and this additionally in an area, which is characterised by an efficient and dense structure, a spreading and a following cascade of activations can occur. However, this range is limited, as no following signals would occur.

An activation cascade triggered in this way has two effects. For one thing, it is asynchronous, that is, it would be detached from all other activation patterns, which would make it both a disturbance and an influencing factor. This difference would depend on whether this would lead to a temporary association or convergence with other neuronal activities, resulting in a one-off change in a connection. On the other hand, it is new: it is new because it would be a cascade of activation based on no other elements.

This novelty is interesting insofar as it would be based on previous weightings in the propagations, but not in the connection with others and thus with Hebb's learning rule: "Neurons that fire together wire together" (Rösler 2011, p. 31). When a connection is established by a random combination of this random cascade and a current activity pattern, it can create a new association. Because the propagation range of such a random cascade also depends on the linking efficiency in this network, it should be clear that in a network of expertise this can extend further than usual, even if it is only one neuron further. In this context, such a one-off activation should not yet be overstated. However, one can further imagine that expertise means an intensive examination of a matter, and that this therefore also takes place sustainably, and that these areas are used more often and in a similar way, so that this accidental activation can occur more often and then lead to a pathogenesis over time. If, in such a situation, the mediator's attention is reduced, that is, no longer has a controlling or contrasting effect, such a developed idea can penetrate the consciousness and appear as a sudden inspiration. A similar thing could have happened with Kekulé, whose question always revolved around the structure of benzene, when he, dozing in front of the fire, saw the tongues of fire reaching into himself. Perhaps triggered by the fire, but perhaps also by a randomly triggered activation subsequently related to the flames, he remembered the Ouroboros and thus found the analogy to the benzene ring.

Another limitation in this matter should be mentioned: (auto)associative networks "complete incomplete initial samples, generalize and are immune to injuries. They also implicitly acquire the prototype of a stimulus set" (Rösler 2011, p. 37). Because the network structure expresses relationships among each other through associative networks, a displacement of these is possible only to a limited extent. However, the formation of new compounds is not uniformly limited in this context, as this is not yet pronounced in this sense. The same was already evident in the way neurogenesis functions and integrates.

This theoretical thought experiment can now be related to the experiment carried out by Guttmann, in which he played physically constant tones to test persons over a longer period of time; the subjects perceived them differently, which could also be shown in the EEG (cf. Guttmann 2016, p. 1f.). The characteristics derived from the basis of this thought experiment would suggest that the frequent and always same stimulation of the same areas has an asynchronous effect. As a cause, this would imply the different perception by a "fatigue" by ATP deficiency in the respective neuron associations, at least in parts of these. This ATP deficiency would therefore lead to short(est)-term problems in restoring the ion balance necessary for the resting membrane potential and thus to distortions.

If this thought experiment and its theoretical implications were consistent with the reference of Guttmann's experiment, it would serve as an analogy, as in a network of dense similar input signals, as in problem solving, where one gets stuck and a sudden shift can occur through "fatigue." If the attention in this situation is not limited because it is in a dozing or similar state, then a new "something" could come into consciousness more easily.

The effect of the random triggering of action potential may well be a phenomenon that occurs frequently. In most cases it has little or no meaning. Only when this leads to an increased probability of usable asynchronicity through the connection with intensively used areas, will chance be able to influence creativity as a moment. Thus, coincidence in this sense certainly does justice to its name, if it leads to a coincidence of local membrane potentials. This coincidence leads to an asynchronous triggering of an action potential under reduced attention conditions and can thus bring something new up to the level of consciousness.

There are, as already noted in Sect. 7.1.7, many other variants of how a person's neural networks can be changed by chance. Stroke patients, for example, are particularly tragic and usually associated with limitations, just as in very rare cases special forms of abilities can occur. Here again the savants Derek Amato or Jason Padgett are to be mentioned, those persons equipped with a savant gift whose abilities might have been released by a strong vibration of the head. Because these special cases cannot be presented here in detail, however, only the aspect of random activation has been considered for randomness in the course of the investigation results.

References

Books and Journals

Anderson, J. R. (1996a). *Cognitive psychology. An introduction* (2nd ed.). Heidelberg: Spectrum Academic Publishing House Ltd.

Anderson, J. R. (1996b). ACT. A simple theory of complex cognition. *American Psychologist, 51*(4), 355–365.

Anderson, J. R., Bothell, D., Byrne, M. D., Douglass, S., Lebiere, C., & Qin, Y. (2004). An integrated theory of the mind. *Psychological Review, 111*(4), 1036–1060.

Andreae, L. C., & Burrone, J. (2015). Spontaneous neurotransmitter release shapes dendritic arbors via long-range activation of NMDA receptors. *Cell Report, 10*(6), 873–882.

Andreasen, N. C. (2005). *The creating brain. The neuroscience of genius*. New York/Washington, DC: Dana Press.

Athanasopoulos, P., Bylund, E., Montero-Melis, G., Damjanovic, L., Schartner, A., Kibbe, A., ... & Thierry, G. (2015). Two languages, two minds: Flexible cognitive processing driven by language of operation. *Psychological Science, 26*(4), 518–526.

Baddeley, A. (2003). Working memory: looking back and looking forward. *Nature Reviews Neuroscience, 4*, 829–839.

Bouhuys, A. L., Bloem, G. M., & Groothius, T. G. G. (1995). Induction of depressed and elated mood by music influences the perception of facial emotional expression in healthy subjects. *Journal of Affective Disorders, 33*(4), 215–226.

Bower, G. H. (1981). Mood and memory. *American Psychologist, 36*(2), 129–148.

Brozinsky-Schwabe, E. (2011). *Intercultural Communication*. Wiesbaden: Publishing House for Social Sciences.

Cahn, R. B., & Polich, J. (2006). Meditation states and traits: EEG, ERG, and neuroimaging studies. *Psychological Bulletin, 132*(2), 180–211.

Carson, S. (2010). *Your creative brain*. San Francisco: Jossey-Bass, A Wiley Imprint.

Chepenik, L. G., Cornew, L. A., & Farah, M. J. (2007). The influence of sad mood on cognition. *Emotion, 7*(4), 802–811.

Cook, I. A., Abrams, M., & Leuchter, A. F. (2016). Trigeminal nerve stimulation for comorbid posttraumatic stress disorder and major depressive disorder. *Neuromodulation, 19*(3), 299–305.

Cryan, J. F., Bravo, J. A., Forsythe, P., Chew, M. V., Escaravage, E., Savignac, H. M., ... & Dinan, T. G. (2011). Ingestion of Lactobacillus strain regulates emotional behavior and central GABA receptor expression in a mouse via the vagus nerve. *Proceedings of the National Academy of Sciences, 108*(38), 16050–16055.

Davies, M. (2003). The role of $GABA_A$ receptors in mediating the effects of alcohol in the central nervous system. *Journal of Psychiatry & Neuroscience, 28*(4), 63–74.

De Bellis, M. D., Clark, D. B., Beers, S. R., Soloff, P. H., Boring, A. M., Hall, J., ... & Keshavan, M. S. (2000). Hippocampal volume in adolescent-onset alcohol use disorders. *American Journal of Psychiatry, 157*(5), 737–744.

Ding, X., Tang, Y. Y., Cao, C., Deng, Y., Wang, Y., Xin, X., & Posner, M. I. (2015). Short-term meditation modulates brain activity of insight evoked with solution cue. *Social Cognitive and Affective Neuroscience, 10*, 43–49.

Ekman, P., & Davidson, R. J. (1993). Voluntary smiling changes regional brain activity. *Psychological Science, 4*, 342–345.

Fademrecht, L., Bülthoff, I., & de la Rosa, S. (2016). Action recognition in the visual periphery. *Journal of Vision, 16*(3), 33.

Franke, A. G., & Lieb, K. (2010). Pharmakologisches Neuroenhancement und „Hirndoping". *Bundesgesundheitsblatt - Gesundheitsforschung - Gesundheitsschutz*, 53(8), 853–860.

Gerrig, R. J. & Zimbardo, P. G. (2007). *Psychology* (18th updated edition). Munich: Pearson Education Germany GmbH.

Goodwin, D. W., Crane, J. B., & Guze, S. B. (1969). Phenomenological aspects of the alcoholic "blackout". *The British Journal of Psychiatry, 115*(526), 1033–1038.

Goodwin, D. W., Powell, B., Bremer, D., Hoine, H., & Stern, J. (1969). Alcohol and recall: State-dependent effects in man. *Science, 163*(3873), 1358–1360.

Grossman, P., Niemann, L., Schmidt, S., & Walach, H. (2004). Mindfulness-based stress reduction and health benefits: A meta-analysis. *Journal of Psychosomatic Research, 57*, 35–43.

Guttmann, G. (2016). Erfinden statt entdecken. In N. Holzenthal (Ed.), *Constructing reality/ Realität konstruierend* (pp. 25–32). Frankfurt am Main: Peter Lang Publishers.

Häusser, M., Raman, I. M., Otis, T., Smith, S. L., Nelson, A., Du Lac, S., ... & Yarom, Y. (2004). The beat goes on: spontaneous firing in mammalian neuronal microcircuits. *Journal of Neuroscience, 24*(42), 9215–9219.

Hirsch, J., Kim, K. H., Relkin, N. R., & Lee, K. M. (1997). Distinct cortical areas associated with native and second languages. *Nature, 388*, 171–174.

Ito, T. A., Chiao, K. W., Devine, P. G., Lorig, T. S., & Cacioppo, J. T. (2006). The influence of facial feedback on race bias. *Psychological Science, 17*(3), 256–261.

Jais, A., Solas, M., Backes, H., Chaurasia, B., Kleinridders, A., Theurich, S., ... & Alber, J. (2016). Myeloid-cell-derived VEGF maintains brain glucose uptake and limits cognitive impairment in obesity. *Cell, 165*(4), 882–895.

Ji, D., & Wilson, M. A. (2007). Coordinated memory replay in the visual cortex and hippocampus during sleep. *Nature Neuroscience, 10*(1), 100–107.

Kahle, W. & Frotscher, M. (2005). *Nervensysteme und Sinnesorgane.* In: Taschenatlas Anatomie. 3rd band (9th revised edition). Stuttgart. Thieme Publishers.

Kempermann, G., Chesler, E. J., Lu, L., Williams, R. W., & Gage, F. H. (2006). Natural variation and genetic covariance in adult hippocampal neurogenesis. *Proceedings of the National Academy of Sciences, 103*(3), 780–785.

Kennedy, D. (2004). Neuroscience and neuroethics. *Science, 306*(5695), 373.

Lewis, M. L., & Frank, M. C. (2016). The length of words reflects their conceptual complexity. *Cognition, 153*, 182–195.

Maguire, E. A., Woollett, K., & Spiers, H. J. (2006). London taxi drivers and bus drivers: A structural MRI and neuropsychological analysis. *Hippocampus, 16*, 1091–1101.

Majumdar, M., Grossman, P., Dietz-Waschkowski, B., Kersig, S., & Walach, H. (2002). Does mindfulness meditation contribute to health? Outcome evaluation of a German sample. *The Journal of Alternative and Complementary Medicine, 6*, 719–730.

Marchand, W. R. (2014). Neural mechanisms of mindfulness and meditation: evidence from neuroimaging studies. *World Journal of Radiology, 6*(7), 471–479.

Mednick, S. (1962). The associative basis of the creative process. *Psychological Review, 69*(3), 220–232.

Mednick, S. A., & Mednick, M. T. (1967). *Examiner's manual, remote associates test.* Boston: Houghton Mifflin.

Mogge-Grotjahn, H. (2011). *Soziologie: eine Einführung für soziale Berufe.* Freiburg im Breisgau: Lambertus Publishing House.

Navon, D. (1977). Forest before trees: the precedence of global features in visual perception. *Cognitive Psychology, 9*, 353–383.

Pease, A., & Pease, B. (2004). *The definitive book of body language*. New York: Bantam Dell.

Perani, D., Paulesu, E., Galles, N. S., Dupoux, E., Dehaene, S., Bettinardi, V., ... & Mehler, J. (1998). The bilingual brain. Proficiency and age of acquisition of the second language. *Brain: a journal of neurology, 121*, 1841–1852.

Pierucci-Lagha, A., Covault, J., Feinn, R., Nellissery, M., Hernandez-Avila, C., Oncken, C., ... & Kranzler, H. R. (2005). GABRA2 alleles moderate the subjective effects of alcohol, which are attenuated by finasteride. *Neuropsychopharmacology, 30*, 1193–1203.

Prinz, M., Erny, D., de Angelis, A. L. H., Jaitin, D., Wieghofer, P., Staszewski, O., ... & Schwierzeck, V. (2015). Host microbiota constantly control maturation and function of microglia in the CNS. *Nature Neuroscience, 18*, 965–977.

Rieken, B. (2010). Das Analogiedenken und seine Bedeutung für Medizin und Psychotherapie. Potsdam. AGEM (Arbeitsgemeinschaft Ethnomedizin e.V.). *Curare*, 33, 192–199.

Ritter, S. M., Damian, R. I., Simonton, D. K., van Baaren, R. B., Strick, M., Derks, J., & Dijksterhuis, A. (2012). Diversifying experiences enhance cognitive flexibility. *Journal of Experimental Social Psychology, 48*, 961–964.

Rösler, F. (2011). *Psychophysiologie der Kognition. Eine Einführung in die Kognitive Neurowissenschaft.* Heidelberg: Spectrum Academic Publishing House.

Roth, G., & Strüber, N. (2014). *Wie das Gehirn die Seele macht* (5th ed.). Stuttgart: Klett-Cotta Publisher.

Schenk-Danzinger, L. (2006). *Entwicklungspsychologie.* Vienna: G&G Publishing Company. (2nd edition in complete revision by Karl Rieder. Reprint 2011).

Shansky, R. M., & Lipps, J. (2013). Stress-induced cognitive dysfunction: hormone-neurotransmitter interactions in the prefrontal cortex. *Frontiers in Human Neuroscience, 7*, 278–283.

Shors, T. J. (2009). Saving new brain cells. *Scientific American, 300*(3), 47–54.

Siegler, R. S. (2000). Unconscious insights. *Current Directions in Psychological Science, 9*(3), 79–83.

Silbernagl, S. & Despopoulos, A. (2003). *Taschenatlas der Physiologie* (6th corrected edition). Stuttgart: Thieme Publishers.

Silver, I. A., Deas, J., & Erecinska, M. (1997). Ion homeostasis in brain cells: differences in intercellular ion responses to energy limitation between cultured neurons and glial cells. *Neuroscience, 78*(2), 589–601.

Singer, J. A., & Salovey, P. (1988). Mood and memory: evaluating the network theory of affect. *Clinical Psychological Review, 8*(2), 211–251.

Spitzer, M. (2009). *Lernen – Gehirnforschung und die Schule des Lebens.* Heidelberg: Spectrum Academic Publishing House.

Strack, F., Martin, L. L., & Stepper, S. (1988). Inhibiting and facilitating conditions of the human smile: a nonobtrusive test of the facial feedback hypothesis. *Journal of Personality and Social Psychology, 54*(5), 768–777.

Tang, Y. Y. (2011). Mechanism of integrative body-mind training. *Neuroscience Bulletin, 27*(6), 383–388.

Tang, Y. Y., Ma, Y., Fan, Y., Feng, H., Wang, J., Feng, S., ... & Zhang, Y. (2009). Central and autonomic nervous system interaction is altered by short-term meditation. *Proceedings of the National Academy of Sciences, 106*(22), 8865–8870.

Tang, Y. Y., Lu, Q., Geng, X., Stein, E. A., Yang, Y., & Posner, M. I. (2010). Short-term meditation induces white matter changes in the anterior cingulate. *Proceedings of the National Academy of Sciences, 107*(35), 15649–15652.

Watson, D., Clark, L. A., & Tellegen, A. (1988). Development and validation of brief measures of positive and negative affect: the PANAS Scales. *Journal of Personality and Social Psychology, 54*(6), 1063–1070.

Weiller, C., Jüptner, M., Fellows, S., Rijntjes, M., Leonhardt, G., Kiebel, S., ... & Thilmann, A. F. (1996). Brain representation of active and passive movements. *Neuroimage, 4*, 105–110.

Wolf, S. A., Möhle, L., Mattei, D., Heimesaat, M. M., Bereswill, S., Fischer, A., ... & Alutis, M. (2016). Ly6Chi monocytes provide a link between antibiotic-induced changes in gut microbiota and adult hippocampal neurogenesis. *Cell Reports, 15*(9), 1945–1956.

Ylönen, H., Lyytinen, H., Leino, T., Leppäluoto, J., & Kuronen, P. (1997). Heart rate responses to real and simulated BA Hawk MK 51 flight. *Aviation, Space and Environmental Medicine, 68*(7), 601–605.

Yu, S. P., Wang, X. Q., Xiao, A. Y., Sheline, C., Hyrc, K., Yang, A., ... & Goldberg, M. P. (2003). Apoptotic insults impair Na+, K+-ATPase activity as a mechanism of neuronal death mediated by concurrent ATP deficiency and oxidant stress. *Journal of Cell Science, 116*, 2099–2110.

Online Resources

Brodbeck, K. H. (2006). Neue Trends in der Kreativitätsforschung. In *Themenschwerpunkt Kreativität, Kunst und Humor*. BÖP. Psychologie in Österreich, pp. 246-253. http://www.khbrodbeck.homepage.t-online.de/trends.pdf. Accessed 2 June 2016.

Kempermann, G., Fabel, K., Ehninger, D., Babu, H., Leal-Galicia, P., Garthe, A., & Wolf, S. (2010). Why and how physical activity promotes experience-induced brain plasticity. *Frontiers in Neuroscience, 4*, 189. https://doi.org/10.3389/fnins.2010.00189.

Kreilhuber, A. (2014). *Körperliche Auswirkungen von Alkohol*. http://www.netdoktor.at/krankheit/koerperliche-auswirkungen-von-alkohol-6451633. Accessed 25 June 2016.

Piore, A. (2013). When brain damage unlocks the genius within. *Popular Science, 2013*. https://www.popsci.com/science/article/2013-02/when-brain-damage-unlocks-genius-within/ Accessed 21 Nov 2020.

Other

Guttmann, Giselher (2013): Von der Erregungssumme zum kortikalen Gleichspannungspotential. Ein Rückblick nach 100 Jahren. *SFU Forschungsbulletin 1(1)*.

Täuber, M. (2016). *Course to become a qualified neuromental trainer. Seminar units 1–4 from February to June 2016*. Vienna: Institute for Mental Success Strategies.

Chapter 8
Research Findings

Contents

The results of the study refer to two aspects. First, these are the aspects that have already been separated in the theoretical consideration of the model of thought, which could be determined as influencing the emergence of something new, and second, the connection of these aspects within the framework of the experimental reference with the attempt to create an understanding that goes beyond and deepens the previous understanding.

8.1 Results from the Theoretical Derivation

The theoretical derivation of the various elements of the model of thought has shown that various aspects can have an effect on the emergence of something new.

The first aspect mentioned was the perception of stimuli, which is already influenced by the respective sensory organs. In this context, the concept of outside and inside lying, related to the sensory elements, has been extended in such a way that outside is not only thought in the sense that it lies outside the body, but also inside it, so long as this can be expressed as stimulus absorption in the broader sense and thus as entrance into the brain.[1] In this sense, this part was seen as a factor

[1] In this context it should be mentioned that the nervous system is an interconnecting and linking system established from growing cell formations, which probably were tasked to establish a con-

K. Testor, *Cognitive Theoretical Foundations of Creativity*,
https://doi.org/10.1007/978-3-030-61736-3_8

influencing both contextualisation in general and contextualisation limited by attention. It is likely to have an influence on the plurality that is important for neurogenesis.

The second aspects were reduced from the derived perceptions but with various other information-associated chunks, which can be found in the contents of the memory in the sense of a perturbation connected from the experienced input signals and the perception filters/experiences/contextualisations already shifting these. These shifts were seen both through the complexity of the associations, which can be expressed as an isa-pointer, as well as through the development and influence of different networks dependent on the activation. Furthermore, it was shown that chunks cannot be thought of as absolute, but are more complex. In this sense, the level dependency of memory contents also became obvious, because two terms, such as "wolf" and "cat," can have a common content higher in a specific cascade, such as mammal, or also in relation to characteristics discriminating and at the same time unifying, such as claws. Because claws occur in both animals, but are subject to different use, these terms are relevant and activated or directed in a different direction, depending on the observation and reference level. The resulting aspects of changeability by chunks were thus seen in the activation of certain contents, the context as well as the level.

The next aspect was the traces of memory, understood as general relationships between memory contents and are the isa-pointer already mentioned in the chunks. The association of these contents may be related to synaptic changes expressed as LTP/LTD. At the same time, the activation emanating from a chunk is limited in width, so it cannot take place in all categories at the same time. Thus, the active memory track has to be distinguished from the general isa-pointer by the fact that it concerns an activated and thus specific isa-pointer in the respective situation. Because the activation probability can be changed by the LTP/LTD, which is expressed among other points in the efficiency, it was derived that the memory traces are seen as influencing variables in relation to the emergence of something new.

The fourth aspect derived from the theory was the rules of production, the procedural contents of memory that express the change of environmental stimuli. So, these are similar to chunks, but with the difference that they are able to express changes and procedures. Rules include context information as well as a layer structure. Within the context, the emotional colouring was included here, because this

nection between sensor cells and motor cells, so between receptors and effectors, and therefore bridge information transfer over increasing distances, as is the case in the organism *Hydra* (cf. Maturana and Varela 2015, p. 166f.). Through the increasing complexity of possibilities of peculiarities in life in these different areas, interconnecting formation extended and therefore created different kinds of receptors, effectors, and interconnections. In this sense the author wrote about parts as "lying outside," which are an entrance for information into the nervous system respectively into the thoughts in the brain; as well, these refer to the "stimuli created from the outside." Through this term expanding view a broader understanding of processes and relations is expected.

can represent important additional information both in the process and in the chunks to be activated.

In the course of the consideration of the aspect of hierarchical control it was determined that this as a negotiation system of competing contents represents a special kind of production rule. It owed its closer examination to the aspect of lateral inhibition, which is a higher-level application of the all-or-nothing decision already found in neurons. At the same time, in spite of this decision, variants that have been left behind are not yet discarded but can be held in a waiting loop and used later or if the framework conditions change. It was precisely this change in framework conditions, as well as the reduction of influencing aspects of control, such as attention, that was expected to have an effect on the emergence of something new. This change can have an amplifying or inhibiting effect as well as allowing the negotiation processes to take place without influence through reduction. A special influence in this context was the context-related goal-oriented action, which in the schemata stored as scripts allows the context to influence the hierarchical control mechanism.

The consideration of the sixth element referred to the neuroanatomical basics, from which it was deduced by connectivity studies that the connection of all areas seems to exist in principle, that certain areas are also difficult to connect with it, but that neurogenesis could change the connection structures. Furthermore, the changes caused by LTP/LTD were included as factors that change in this sense, because they have a more important role in the new integration of neurons than in already established structures.

Another important aspect has been mentioned here that is already found in the definition of thought: the threshold of consciousness. Many of the points mentioned so far are likely to occur frequently. But to be able to speak of the emergence of something new or even of creativity, it must be a conscious process. Without the knowledge or recognition of the same or at least of becoming conscious, it would not be possible to think the new and would also speak against the definitions of creativity, which includes the term value and thus implicitly includes also consciousness as a prerequisite for evaluation.

Also, from this point of view, the expertise was examined more closely, as it is expressed in a denser and more efficiently networked neural network, which should make the connections more flexible, as well as the crossing of the awareness threshold with related areas is likely to happen more easily.

As the last aspect of the neuroanatomical approaches, chance was mentioned, which found expression both in damage and by extension or in spontaneous activation.

In summary, it was concluded from the theoretical derivation that changes can take place through receptors, contextualisation, attention, level reference, emotional colouring, neurogenesis, and LTP/LTD, through the activated contents—both chunks and production rules—as well as through inhibitory attention, through strengthening attention, through reduced attention, intensive introspection as a particularly pronounced focusing of attention isolated from external influences, lateral disinhibition, through expertise and through the necessity of crossing the threshold of consciousness. Finally, the possibility of chance was mentioned and suspected. Thus, the theoretical derivation is as follows:

- Change in stimulus
- Reference level
- Traces of memory
- Activated contents
- Production rules
- Emotional colouration
- Inhibitory attention
- Increasing attention
- Reduced attention
- Intensive introspection
- Reduction of control mechanisms
- LTP/LTD
- Neurogenesis
- Expertise
- Fortuity

Through the reference of the model of thinking with five experiments and the hermeneutic circle as a method of the humanities and finally the theoretical elaboration of the element of chance, represented by the spontaneous or random activation, parallel to the theoretical elaborations became apparent. These findings are examined in more detail in the next sections.

8.2 Results from the Experimental Reference Setting

The experiments have served to establish a reference from these results that can be linked to the model of thinking and should thereby be able to show from a theoretical understanding whether and which aspects could have an influence on the result or to what extent these could be used as a basis for the emergence of something new.

8.2.1 Results of Language as Context Information

The study of multilingual test persons showed that a neutral short video is perceived and described differently by the respective language used. The change of language results in a change of perspective which, to put it bluntly, even brings to light different views of the world: an actionist as opposed to a holistic view. The distinction between the two, derived from linguistics, referred only to the grammatical existence of an expression of the event taking place at a given moment, the present progressive, which exists in English.

The investigation has shown that this grammatical information is to be seen as part of the chunks, which can thus be understood as a rule of production expressed in an isa-pointer. Production rules show how chunks change in relation to the environment. This aspect, already explained in Chap. 7.1.4 Production rules, is thus part of the understanding of the world, because the changing environment can be brought back to action through use in this construct, which is expressed through linguistic regulation, as it was realised in this experiment in describing the situation. The grammatical component can therefore be seen as a rule of production which, when changed, leads to other effects.

Another part of the investigation is the similarity and at the same time the difference of linguistic concepts. The example of "church" and "Kirche", words which are likely to have a very close core of terminology, nevertheless have substantial distinctions in the somewhat broader associations, which are experience dependent. Thus, it is precisely between Austrian Catholic churches and English Protestant churches that not only the underlying religion is different, but also the architecture, style, and history. The change in the use of the term thus leads not only to a change in language, but also to a culturally and experientially influenced change in the "word cloud", that is, the basis and orientation of associative representation. From this shift further different scripts result, as for example in the two churches mentioned, the behaviour and the procedures partly diverge. It is therefore necessary to think along with a changed "event cloud".

The next interesting part of this study was activities based on the limitation of working memory capacities. It was shown that if a second foreign language is present, this language is used as a buffer when two tasks concerning the linguistic areas have to be processed, the monotonous counting and the processing of the visual input. This monotonous counting, which nevertheless used up the working memory in its central processing capacity, made demand of the language that was used for counting. The visual impressions, however, could be processed in the other and became so influential in the subsequent description that, despite the fact that the working language remained the same, this was expressed in a shift in the tendency of the description.

As a derivation, two thoughts concerning attention can be pursued. On the one hand, it is possible through monotonous, nonautomated routines that shifts in the neuronal processes can take place in the background, which result from the utilisation of the control mechanisms of the nonautomated routine treatment. On the other hand, through the subsequent attention to the resulting newness, it is possible to enforce previously weak aspects or rarely penetrated contents through the intensifying effect of the attention in the contrast, so that this could be carried beyond the threshold of consciousness.

In summary, the context shifts caused by the language changes both in chunks and in production rules, the turning away from the normal that can be influenced by attention, the blockade of attention by nonautomated routines as well as the attention paid after such processes to reinforce the possibly existing new, are to be mentioned.

8.2.2 Results from Irritations

In the irritations caused by movement in virtual reality, increased cognitive flexibility was shown. Irritations change them. In the investigation, the model of thought tried to find an explanation for this. It was also shown that by looking together at experiments on the closeness to reality of virtual reality and real experience as well as the neuronal activity in imagining and experiencing, the processes between real and simulated or imagined are very similar and that these can therefore be regarded as a special part of reality if an active component is included.

Based on the fact that the events depicted in virtual reality can be assumed to be extensive in the chunks because of the relative frequency of experience in everyday life, a stable basis was found in the experience and thus used as fertile ground for irritation. In this context, it was also assumed that this applies to the related production rules and that very clear scripts can be expected overall. If a bottle falls off a table, there should be less room for the scripts than there would be for the speed of movement in space. The experience gained in all areas, which was considered to be extensive, also led to the assumption that a correspondingly high density of cross-linking and the associated established propagation paths for the respective activity patterns are present. This situation is likely to be similar to an expertise for the narrowly defined area of the respective events. Consequently, it was to be expected that the priming effects would be prepared by the expected effects and that the corresponding action cascades could be activated more easily. However, the events experienced in the abnormal situation caused irritation. The inner, expected, prepared element differed from the outer activity pattern. This divergence may have resulted in a greater activation span, which was positively reflected in the breadth of responses found in the Unusual Use Tasks.

This divergence was expressed above all in the distinction between active and passive irritations by the fact that the active elements may have coactivated the basal ganglia, which is why in this area the mentioned activation span would be wider than in the passive form, and thus more areas could be available preactivated for the subsequent task.

Because the irritation, which is a particularly pronounced form of rarely used information and therefore uses little to hardly linked paths, activates areas that otherwise did not take place in this form, additional traces of memory are created. These, however, also tie in with established areas or, through relative simultaneity, lead to linking tendencies between the new and established areas. In addition, this is beneficial for the creativity test, as the after-radiation of these areas, which is influenced by the early phase of LTP/LTD, is expressed in an easier excitability.

However, this after-radiation of the memory traces may be limited in its stimulating effect, because in the second experiment two successive tests showed a different characteristic. On the basis of the model of thinking, this can be understood as meaning that the openness and parallelism generated by the irritation was strongly present for the first test. The decisions taken in the first test and the restrictions placed on certain categories of thinking in relation to the points asked for also created a limitation in the categorisation option for the subsequent test.

The irritations had also led to the expectation of completely new applications of production rules from the postponed usefulness analysis resulting from the uncertainty. This change is expressed in a reduced control performance, that is, lateral disinhibition, which thus enables broader negotiation. In this context, it should also be mentioned that the experiments were effective both in a concretely specific schema disorder and in a generally complex situation, which is why level independence is expressed here. This, too, is an indicator for the reduction of hierarchical control.

In summary, the shifting of chunks, scripts, and production rules can be cited here as the basis for the emergence of something new. In addition, the changes of the memory traces come to bear through these, which were not only activated differently starting from the shifted elements, but through priming led to a broader activation whether the discrepancy was inside or outside, which was intensified when the basal ganglia were included in the experiment through active integration of the volunteer. Finally, there are the shifts in the rules of production of hierarchical control caused by the irritation, which were eliminated or at least reduced as a limiting factor for the emergence of something new.

8.2.3 Results from the Investigation of the Influence of the Body on Thinking

In the investigation of the body as an influence on thinking, the distinction between the internal and external parts of thought was strained at the beginning. This distinction is important because the signals resulting from the body posture according to the external understanding of the receptors represent an external aspect, as does the facial expression of "genuine" smiling, which is strongly associated with an emotion, an internal aspect. These internal and external parts of thought have a mutual influence that depends on various factors. Emotion is a specific influencing factor, which can express itself both culturally and universally in certain postures or facial expressions. The conclusions derived from the connectivity studies are that although the areas responsible for the emotional manifestations can be effective in influencing the areas responsible for the cognition, some speak of rationality, inversely this is not the case; it is all the more interesting that the change of the body seems to have an effect in a change of emotion. This idea even goes so far that depressive symptoms can be alleviated by regular specific activation, which could be attributed to changes in the activated chunks and memory traces activated differently by them, as well as to changes caused by LTP/LTD.

The connection between facial expression and emotion is a specific one. The typical expression for an emotion such as joy of a "genuine" smile is distinguished by a specific use of several facial muscles. Only the use very close to the emotional implementation of the smile leads to similar neuronal activity patterns.

Other studies suggest that the emotions are expressed as a special part of the isa-pointer of the chunks. Because the “emotion patterns” created by the altered posture activate a different sphere of action, both chunks and the memory traces and production rules that emanate from them are altered. However, the limitation of this effectiveness is likely to be limited to the recognition of faces and emotional words. The extent to which this is valid as a restriction cannot be precisely stated here, because the very concept of the emotional word as a construct is so individual that the effort to do so can only be made in different studies. Nevertheless, it can be said that there is a connection and that it is subject to limitation.

This influence of the emotion on the isa-pointer of certain chunks leads to a change in negotiation processes from a shift in the isa-pointer. The change of emotions can mean these act as amplifiers on different ranges and lead thereby to changes in the evaluation, whereby new can develop.

In summary, it can be said that the body can change the emotional colouration, which affects chunks, rules of production, and traces of memory as well as the negotiation processes and can thus produce altered activity patterns.

8.2.4 Results of Alcohol as Context Influence

The investigation of alcohol as a context influence aimed primarily at the idea that a reference settlement with the influence of various stimulants on neuronal processes could be considered with the model of thinking. In addition to the harmful effects of alcohol, the influence of alcohol on the development of something new as well as the change in processes and emotional colouring has also been shown.

At the beginning, the similarity of the influence of alcohol with the emotional colouration was assumed, which could express itself as a special part of chunks and thus as part of the “word cloud”. However, this was quickly recognised as unlikely by the consideration of the word association test. Rather, alcohol acts as an influential element in changing the use of memory traces. It is not an isa-pointer, but acts on the GABA receptors, which are important for the inhibitory synapse circuits. Because it docks to the receptors instead of GABA and thus blocks it, the effect of inhibition is reduced and leads to suppressed spreading compared to the normal state.

Because in the normal state certain memory contents are meaningfully inhibited and thus usually barely occur, rare contents will come by this influence more strongly to the carrying than otherwise; this is also part of the shifting effect of alcohol.

However, alcohol still exerts an influence on the emotional colouring, which can be linked to the increased activation associated with the reward centres. Thus, different chunks are activated by the influence of alcohol, because these are activated more easily by the emotional colouring, similar to the experiment with facial expressions. However, this is the side effect that explains the proximity to the assumption just described and not the direct effect of the alcohol. In this sense, those emotional shifts towards melancholy or other emotions should be considered separately.

In addition, inhibition of the inhibitory mechanisms makes it possible to reduce hierarchical control, which can manifest itself in lateral disinhibition and thus increased creativity. Thus, the disinhibition also leads to the reduction of hierarchical control.

In this sense, the reduction of attention is also relevant, which as a limiting factor can partly reduce the restriction of the emergence of new things. Also, in this state of reduced attention it is more difficult to learn new things.

Over a long period of time, the decrease in brain plasticity induced by reduced neurogenesis, which should lead to a decrease in creativity, has a negative effect, as does the change in the use of memory traces caused by LTP, which will become partly dependent on the consumption of alcohol.

In summary, it can be said that alcohol primarily causes a shift in the traces of memory because of its GABA receptor-blocking effect. This shift also leads to the activation of rarely used content, as this would normally be subject to an inhibitory influence. Furthermore, a reduction of the hierarchical control, which also affects the reduced focusing ability of attention, is generated by this effect and can also bring about a different level reference. Ultimately, the influence on different areas affecting emotions also changes the emotional colouring, leading to a variety of different activations of chunks. In addition, the consumption of alcohol is associated with complex and extensive negative effects, which in the long run lead to a reduction in the quality of life as well as a reduction in the ability to create something new.

8.2.5 Results of the Meditation Study

The meditation technique used for the experiment was characterised by the fact that it can be understood both through the detachment created by physical focusing and subsequently through a state aimed at acceptance and relaxed alertness. In addition, this technique can be learned quickly with relatively little previous knowledge.

The first part of the exercise directs the attention to unnoticed areas of the body, thus lifting an otherwise unconscious perception above the threshold of consciousness through the strengthening contrast. At the same time, rarely consciously used chunks are activated and with them the traces of memory emanating from them. Furthermore, other production rules are more likely to be activated than in the normal state. This shift increases synaptic efficiency through repetition, that is, through LTP, and the associated reduction of stress increases learning performance.

In the second aspect, relative lack of evaluation, the aim is to reduce emotional influences. This change should take place similarly to the experiment with the smile about physical influences and reduces the isa-pointer of the emotional colouration. Furthermore, this state aims at a change in hierarchical control, which leads to an increase in introspection and self-perception as well as emotion control through an increased activation of the insula and the superior temporal gyrus. This shift should lead to greater openness to new ideas.

Attention is also influenced: this is brought into a defocused state by relaxed wakefulness, which changes the negotiation processes otherwise influenced by lateral inhibition. The attention should be in an oscillating state, allowing several areas to be covered. Through these two approaches, an additional change in the level reference is carried out, for example, in the physical changes from the macro to the micro level.

Furthermore, through the processes that are changed through meditation, the dopaminergic system becomes activated similar to a flow experience, so that the neuronal changes occurring in these processes are better retained and additionally create the preactivation relevant for the occurrence of *Aha!* experiences, so that crossing the threshold of consciousness of such experiences becomes more likely.

In summary, it can be stated that meditation can bring about the acceptance and thus the reduction of hierarchical systems, as well as the resulting defocusing of attention. In addition, chunks, memory traces, and production rules are used in a new way by the introductory focus on otherwise rarely used areas. With sustained exercise, changes by both the dopaminergic system, the LTP, as well as by the improved integration capacity through stress reduction through adult neurogenesis, become effective in the long term. This change creates an increased openness, which in turn has a positive effect on the emergence of something new and the acceptance mentioned at the beginning.

8.2.6 Results of the Hermeneutic Circle as a Creative Technique

In its process, the hermeneutic circle relies on prior understanding and the associated influence on perception, which thus creates a change in the prior understanding through the new. Based on the results of this theoretical experiment, the chunks and related production rules are to be understood as the basis that constitutes the prior understanding. To be able to integrate completely new content, analogies were used to activate and link the content closest to the previous understanding. This content can be expressed metaphorically in images as well as purely through specific sounds or other forms of proximity. At the same time, this part is individual based, as it depends on experience.

The change in this case would take place through the presence of the memory contents activated by analogy, which would then be able to activate new traces of memory and thus create specific influences on networking and integration, depending on their characteristics and topic.

The prerequisites are openness and reduced contrast from reduced attention. If the completely new can cross the threshold of consciousness, attention is important as a strengthening mechanism for a sustainable change of chunks and production rules.

8.2.7 *Results of the Thought Experiment Fortuity as the Basis for New Things*

The basis of the new in the thought experiment was the random triggering of an action potential. It turned out that this is conceivable in principle. As a prerequisite, an intensive use of an area is necessary to lead either to an extended refractory phase or to an additional triggering. The resulting asynchronicity has a limited range and is small in influence. Only the presence of transfer-efficient structures, as is the case, for example, with regard to expertise, increases the probability that such asynchronicity leads to an influence. This influence can be expressed in the displacement or initiation of a random cascade, just as it can lead to a suppression through the prolonged refractory phase. The latter aspect is less effective, however, because one of the characteristics of (auto)associative networks is their relative injury immunity and thus the failure of a neuron would have no real meaning.

If, however, a random activation does not only take place in the efficient structure, but also through a subliminal preactivation for a topic important to the person or an important question, whether ascribed importance or not, it would be possible to cross the threshold of consciousness. This move is all the easier when attention is reduced and so are the hierarchical controls, so that a weak activity pattern can prevail.

In the case of important topics, especially those that move in the fields of expertise, the occurrence of such random activations is more quickly able to create a path, whereby in these areas the probability increases that something new is created by such an effect. The process of neurogenesis in this area must also be taken into consideration, because the integration of new neurons can have a similar effect, although they are not integrated purely by chance. The effect of chance can thus be seen in both the area of the chunks and the associated memory traces or production rules.

In summary, it can be said that the influence of chance can be limited by the encounter of efficient networks with reduced control mechanisms, and that over time it can create the crossing of the threshold of consciousness through facilitation and integration and thus be of significance for the emergence of something new.

8.3 The Aspects and Their Effectiveness

Overall, the experiments have shown that the aspects of the model of thinking derived from the theoretical aspects can be coherently related to the emergence of something new. Table 8.1 gives an overview of the respective numbers.

Table 8.1 Aspects of the new: number in relation to the frequency in the studies

Aspect of the new	Number
Change in stimulus → change in association	3
Change in the reference level → discriminatory cascade	5
Traces of memory → associative strengths	6
Rarely used memory contents → new starting point	4
Production rules → re-interpretation/change of the approaching	4
Emotional colouring as specific context information → production rules	4
Reduced attention → reduction of contrast	3
Attention as a strengthening control mechanism → raise weak signals above the threshold of consciousness	5
Intensive introspection → stronger decoupling of context information	1
Reduction of control mechanisms → lateral disinhibition and consequently alternative negotiation situations	5
Neurogenesis → change through learning/experience and integration of new products and services	2
LTP/LTD → change in conductivity and efficiency	4
Expertise → high density and conductivity → easier excitatory	3
Random triggering → random triggering and restructuring	1
Chunk change → change of the activated contents and thus their associations	7
Context change	2

The frequencies listed here are quantitative frequencies taken from the results, which could be mentioned once for each experiment in the respective areas. Thus, for example, the random triggering occurred only once, as this occurred only in the thought experiment of chance. The intensive introspection was only once present as part of the meditation technique. The two aspects mentioned in italics were not explicitly derived from the theoretical derivation, but they were explicitly mentioned in the investigation. These aspects are mentioned here for the sake of completeness but are very closely connected with the concepts of memory traces as well as the changing of the level of reference. This result is to be made more comprehensible by the detailed breakdown, which will no longer be carried out experimentally, but in relation to the individual aspect.

8.3.1 Change in Stimulus

The change in stimulus was found in the experiment on irritations. The change of familiar environmental stimuli and their sequence produced changes in the presentation of stimuli. A special example would be the suitcase, which becomes smaller when approaching and larger when removing.

In the experiment with alcohol, the change in stimulus was seen through the blockade of the GABA receptors, which has such an effect that stimuli are perceived differently through the reduction of the inhibitory synapses.

In theoretical random activation, it was assumed that if such an activation takes place, depending on the spatial location of the occurrence, it can cause a change between stimulus and association. This aspect has also been adopted for other areas where the same applies mutatis mutandis. In the following list, these aspects are only mentioned with the abbreviation TRACT (Theoretical Random ACTivation).

8.3.2 Change in the Reference Level

In the experiment of bilinguality and the contextualisation derived from it through language, a change in the reference level was shown. This change had the effect that working memory was blocked by a nonautomated routine task and thus the second language was used in the background to process incoming stimuli. By omitting the routine task, this shift was reversed, and the change in reference had an effect on the verbally used language level in the perspective associated with the second language.

In the experiment of smiling as part of body language, the conscious ingestion of a "genuine" smile resulted in a change in the emotion, which affected the internal and external parts of the thought in a change in the reference level emotion and made a change possible.

A similar effect was caused by the use of alcohol. On the one hand, a change in the emotional colouring was achieved and thus a change in reference analogous to the smile. On the other hand, by blocking the GABA receptors, a change in the reference level was created by limiting the influence of different control levels.

In meditation, the change of the reference level was achieved by focusing from the macro level to the micro level of physical aspects, as well as by allowing otherwise unperceived emotions or thoughts to take place in a depth that was not otherwise present.

In TRACT, the shift in focus associated with the subliminal preactivation has been conceived as an element that simultaneously supports the creation of new associations. This development is of greater importance in areas of expertise as it allows easier triggering through more efficient networks.

8.3.3 Traces of Memory

In the language experiment, the memory traces have produced different associations through the change of language, although these also represent different empirical values in this context. Thus, the easier activatability of more frequently experienced compounds, conceived under the term priming, is irritable by a change and thus possible for the creation of something new.

This priming, which also results from the scripts, was also important as an aspect in the experiment on irritation. By preactivating the frequent associations and the then occurring abnormal processes, the activity pattern was extended by a nonexistent self-similarity and thus caused a broader preactivation, which had a positive effect on the creativity tests. Important here is the active involvement to achieve the activation of the basal ganglia and thus to generate a larger activation range. It may also be important to note that the extension of the activation span only works over a certain period of time and, by holding a creativity test with the specific question it poses, seems to create a limitation in the subsequent phase.

In the experiment with alcohol, the blockade of the GABA receptors caused a change in the spread of the memory traces, which had the effect that these were qualitatively and quantitatively far removed from those in the sober state.

In meditation, a conscious schematic observation produced a change in the traces of memory, which not only spread out from new or rarely used chunks, but also took different paths through changes in attention and hierarchical control mechanisms.

For the hermeneutic circle, the traces of thought have an important role in the prior understanding, because they form a preferred chain of associations for this, and also because they are significant for the analogy formation of something completely new. In this case, they move from the nearest found chunk into different paths influenced by different associations. Thus, the traces sometimes shape and change new tracks. They can also be found in the TRACT by randomly activated pathways.

8.3.4 Rarely Used Memory Contents

This part of the creation of something new are chunks that are (at least consciously) subject to rare use. Therefore, the blockade of the GABA receptors is essential in the experiment with alcohol by shifting its memory traces. Through this blockade, those areas which are normally reduced and affected by inhibition are no longer activated, inhibited, and thus accessible.

A similar effect is created in meditation, which is expressed by concentrating on areas that are rarely in the focus of attention, thus bringing memory content in the background to the fore. Under this aspect also the more intensive observation of a motion, which is understood as carried out in higher detail resolution, is to be seen, which is to be found thereby both in the reference level shift and in influences as rarely used memory contents.

In the hermeneutic circle, new and seldom used memory contents are created by connecting completely new things. At the same time, these can be found either in very frequently used or in very rarely used connections. Thus, this aspect is also relevant in this humanistic method. For the TRACT, the rare memory contents are to be understood analogously to those listed here.

8.3.5 Production Rules

In the language experiment, the grammatical, that is, the linguistic, rules are understood as rules of production. These rules are the changes of chunks experienced from the environment, which express themselves on the one hand in the application, but with it also in the effect on perception and recognition. By changing them, not only the grammar but also the interpretation of the event can be changed.

In the experiment on irritations, the active involvement of the volunteers activates the routines and production rules stored in the basal ganglia, which are stimulated by the subsequent changes in reality and thus lose their usefulness as a point of reference. Completely different processes make it possible to postpone the activation of atypical production rules from the context, just as this occurs if previously unsuitable production rules were linked to the prepared production rule in terms of product composition.

Through meditation, a shift in the application of production rules is created in the area of production rules through changed processes, just as those less conscious can be brought to the fore through this, affecting the associated activation cascades as well as the possible production combinations. The TRACT can also be seen to have an influence here.

8.3.6 Emotional Colouration

Emotional colouration as a special isa-pointer became apparent in the experiment on body language. By changing the physical signal inputs, it was possible to show an influence on the emotional expressions and thus on the associated chunks, production rules, and memory traces as well as the negotiation processes. The limitation to facial recognition and the linguistic range of emotional words was shown, although this leads in its subjective differentiation in the sense of constructivism to different widths in the division.

A similar situation was observed in the area of alcohol consumption: this leads to a change in the emotional basis and thus to a change in the colouring of the same, which is expressed in both the interpretation of moods and the memory of emotionally charged images.

In meditation, by increasing the dopaminergic effects, a flow-like state can be stimulated, which is sometimes expressed in an increased willingness to learn. Furthermore, a reduction of emotional influences is created by the relative detachment, which sometimes shows itself in an increasing openness for the new and the allowing of something originally not one's own. Similarly, TRACT can also have an effect here.

8.3.7 Reduced Attention

In the language experiment, attention was drawn to a nonautomated routine task and thus reduced for other areas. As a result, the use of the second language for the processing of visual information was no longer inhibited and could be incorporated into the processing.

In the case of alcohol consumption, attention is reduced and, in the case of larger quantities, the focus is influenced at the expense of memory and learning performance, which results in a reduction in contrast and thus to changed negotiation processes that can create something new.

For the emergence of the new in the hermeneutic circle, a reduction of attention to the point of dozing is advantageous, because the omission of contrast enhancement to the normal and familiar knowledge can bring a weaker one forward from the background.

8.3.8 Attention as a Strengthening Control Mechanism

By consciously focusing on a language other than the one normally used, memory content associated with it can be strengthened and, supported by attention, integrated more sustainably. In this context, the connection with the otherwise rare content is of essential importance, as is the influencing of memory traces.

What has not yet been experienced only becomes conscious through the focusing of attention and can therefore be integrated. Similar to experiments in which something unusual is suspected from the very first appearance, but is only seen consciously after 8 to 12 performances,[2] attention can be understood as a mechanism that raises the unusual and the new to consciousness.

The effect is similar in meditation. The body signals or thoughts and emotions amplified by the focus of attention allow these to assert themselves against otherwise dominant ones and thus prepare the ground for a change.

The same applies for the hermeneutic circle. Especially after the loss of attention and the implementation of new ideas from unfamiliar areas, the focus of attention on these new ideas can lead to their consolidation and improved integration. It is therefore an effect that is particularly important in the creativity cycle in the verification phase.

---

[2] Exemplarily, a study is mentioned in which the tested persons had to look at a computer screen, which shortly turned black after a few seconds randomly to simulate a blink. The pictures seen on the screen before and after the "blinks" were nearly identical, but for one single difference. This difference was turned on and off alternately with every "blink." Despite this, the probands usually realised a difference after the first change, but it took them several repetitions to figure out what exactly the change was (cf. Rensink et al. 1997, pp. 370, 372).

In the case of TRACT, this area is attested to be important insofar as it can help a subliminal preactivation combined with a new association to cross the threshold of consciousness, whereby the new suddenly comes to mind and can thus experience a consolidation in the verification phase analogous to the understanding cited in the hermeneutic circle.

8.3.9 Intensive Introspection

Intensive introspection was only found during meditation, as this is a special form of concentration. Compared to the comparison group, increased activity was observed in the insula and the superior temporal gyrus. For these areas, in addition to the increase in introspection, self-perception and emotion control are also associated. This type of introspection should have the effect mentioned by Arne Dietrich, which suppresses incoming signals from other areas and thus manages to work primarily with the available information (cf. Carson 2010, p. 168).

8.3.10 Reduction of Hierarchical Control

The reduction of hierarchical control, which among other changes results in lateral disinhibition, becomes effective in a similar way to the change in production rules. Thus, in the experiment on irritation by changed everyday events, a reduction of the influences of higher areas is to be expected, because these cause a changed reference to the lower areas. In addition, the influence of top-down motivation can have different effects.

In the case of alcohol, the reduction of hierarchical control is demonstrated by the fact that the inhibitory circuits necessary for this are blocked by alcohol and thus no longer work in the effectiveness existing in the sober state. Changes in this area, which can be expressed in new ways, can follow.

In meditation, contrast is reduced by the state of defocused attention, and oscillating phases can occur in the negotiation processes, in which both the processes themselves and the attention affecting them are an influence.

It is important for the hermeneutic circle that there is no negative attitude that counteracts the integration of new knowledge and has an inhibiting effect on it.

In TRACT, a reduction in hierarchical control leads to weaker signals that are able to penetrate consciousness more easily with reduced, or even no, negotiation limitation.

8.3.11 Neurogenesis

Neurogenesis, despite its otherwise important role in (re)stabilising neuronal structures, has only a minor effect in the experiment of alcohol as a context influence, because it is disturbed by alcohol consumption. In addition, the situations experienced under the influence of alcohol are integrated with the alcohol-influenced spread of memory traces and chunks, which can suggest the reference settlement occurring through limited neurogenesis is more difficult to establish in an empty state.

In meditation, on the other hand, the reduction of stress leads to an increase in neurogenesis and thus to a better ability to learn and higher plasticity. In addition, this can create an improved networking of the new elements with the existing elements, so that they can be merged into something new in the longer term.

8.3.12 LTP/LTD

The effect of LTP/LTD on the irritation experience is that the memory traces activated for this situation are used more widely: this is accompanied by a better connection in the associative fields, which, if further promoted, can have a lasting effect.

LTP/LTD is likely to have an important role in influencing body input signals in PTSS treatment, as it provides relief shifted by improved synapse efficiency through regular and nonstress-associated stimulation. New pathways are thus strengthened and will probably make it possible to bypass those stress related.

The LTP/LTD is also important in meditation for the sustainability of effectiveness, as it strengthens new paths through different use and novelty.

LTP/LTD is particularly important as a prerequisite for TRACT, as it enables easier triggering in network areas that are already more efficient, but also because it creates easier excitability during a first random temporary adaptation in the early phase, which can be a positive support for repeated activation.

8.3.13 Expertise

Because of the denser and more efficient interconnection, expertise is relevant for the resulting easier activation; this was already relevant in the experiment of irritation, because it resulted in a comprehensive experience for an event, which was gathered from everyday life. From this a rapid and already situation-specific efficient activation is to be expected, which shows up in the irritation as basis for a broader activation span influenced by the irritation.

In meditation there is an increase in expertise over a long period of practice, which is based on better performance in other fields through easier attainment of a defocused focus or also of the other elements mentioned in meditation.

For the thought experiment of TRACT, the expertise is of special importance, because in this area, because of the more efficient networking, there is a higher probability of a random potential becoming effective compared to other areas.

8.3.14 TRACT

Because this was only conducted as a thought experiment, TRACT is not mentioned as an aspect in any other experiment. Taking into account the causes cited in the experiment, however, it can be assumed that an intensive and relatively monotonous activation of the brain — be it by counting in the language experiment or by focusing on a specific area of the body such as smiling or meditating, or also by focusing intensively in the hermeneutic circle to solve an integration problem — can also be assumed to trigger a theoretical random activation that would subsequently have effects.

8.3.15 Chunk and Context Change

The change in chunks was found in all experiments, which is not surprising because to become mentally active, the activation of chunks can always be understood as relevant. At the same time, the experiments triggered changes in the chunks or their use, which made it possible to create new connections. Thus in the language experiment different language clouds as well as isa-pointers influenced by grammar were used, scripts were shifted by the irritations in the second experiment, emotions were identified as isa-pointers for parts of the memory content, chunks otherwise inhibited were activated by alcohol, and new chunks were used in meditation and reference levels were changed; in the hermeneutic circle, they were expanded and reconnected by connecting new chunks, as well as asynchronously activated and associated in TRACT.

The change of context was seen in the context of the change of language in a changed framework of experience and the thus shifted contextualisation. In TRACT, this was enabled by random and asynchronous activation.

This finding shows that such a clear separation of the derived aspects is not so simple, but rather that they are overlapping constructs, which may, however, be related to each other in a similar way of functioning.

8.4 Recapitulation

Although the clear separation between the areas of memory content and their neuronal components as a construct has to experience a restriction in categorisation (to be discussed later), it can be said that aspects of theoretical derivation have been found in the experimental investigation. The changeability of these aspects[3] as an element of creating something new are discussed in the next chapter, Chap. 9.

References

Books and Journals

Carson, S. (2010). *Your creative brain*. San Francisco: Jossey-Bass, A Wiley Imprint.

Maturana, H. R., & Varela, F. J. (2015). *Der Baum der Erkenntnis* (6th ed.). Frankfurt am Main: Fischer Paperback Publisher.

Rensink, R. A., O'Regan, J. K., & Clark, J. J. (1997). To see or not to see: the need for attention to perceive changes in scenes. *Psychological Science, 8*(5), 368–373.

[3] Attention (defocused, focused, reduced, strengthened), chunks, emotional colouring, expertise, memory traces, hierarchical control, intensive introspection, context, LTP/LTD, neurogenesis, production rules, stimulus change, rarely used memory content, TRACT.

Chapter 9
Results Discussion

Contents

In the synopsis of the results of the investigation, the aspects found from the theoretical derivation were reflected in the experimental reference setting. The interpretation associated with these showed at least one correspondence for all elements. However, because the aspects in the demarcation as well as the relation are quite similar among themselves, it can be assumed in the sense of constructivism that the distinctions made by the author are interchangeable in some aspects. Why and to what extent should be the first point of discussion.

When talking about chunks, the isa-pointers connected with them are also included. However, the isa-pointers are the basis of the expansion of memory traces. Because traces of memory are simultaneously expressed in a change from one chunk to the next, this means that consequently even production rules cannot be differentiated sharply. Through the synaptic connection and its modification, the LTP/LTD is brought on board at the same time, as is neurogenesis for integration. In terms of hierarchical control and attention, these can be seen as a special type of production rule that begins to close the circle again. Because expertise, emotional colouring, and the rarely used memory contents are nothing else than parts of the chunks, there would be no new element in this case. In this argumentation, apart from TRACT, which cannot be checked or confirmed, only the context information is missing. Also, this is only a part of the isa-pointer of the chunks, which can also express itself in scripts.

The combination illustrated here, which could at the same time be applied as criticism, clearly shows that the thought constructs used manifest themselves as aids

K. Testor, *Cognitive Theoretical Foundations of Creativity*,
https://doi.org/10.1007/978-3-030-61736-3_9

to understanding. Because subsumed, they can all be reduced to a common denominator: memory content. And as such a differentiated concrete consideration would not be possible, because memory content cannot be examined without abstraction and reference. Thus, these constructs are used to create the beginning of an orientation aid moving blindly in the forest.

Just as constructs generally must be subjected to questioning to gain knowledge in reflection, the reduction to the concept of memory content is too abstract in this area. As with all abstractions, essential aspects of the expression are sometimes subjected to smoothing and are thus no longer represented in real but in a distorted expression. Therefore, the previous terms are admissible to make the different expressions of the memory content more tangible. In the reduction, the author already mentioned an essential definition in this paper, which encompasses these aspects listed here. Thus, the thought was defined as a construct consisting of an inner and an outer part of thought: this is, if one begins to look more closely at the inner part of the thought, an outer one extended by the internal reference. The inner parts of thought arise from the fact that spatial temporal associations can be formed from external stimuli in different areas, up to the individual convergence zones, which can then be further developed by a further self-referentiality. Thus, the development from a language of sound allocations related purely to objects is possible to a language that changes and relates itself as well as the objects, as the word "because", for example, can express. The word because is an expression of a realisation of a relationship, which the person expressing this word must first of all have experienced[1] and created from the environment, to experience realisation through this constructed understanding as an inner part of thought. Through this internal realisation, it is possible to use it as an understanding of reference in the contemplation of the world, and also to express it in the sound attribution associated with the term, whereby it can become relevant as an external part of thought for someone else.

This point of discussion is also a first result. Creativity may present itself as ordinary or exceptional creativity, although this distinction might be subject to the limiting subjective confusion "to consider one's own inventions as discoveries" (Guttmann 2000a, p. 74), as becomes obvious in the definition of the concept of creativity cited by Brodbeck, when, in addition to novelty, the attribution of value must take place. He explains in more detail that this attribution is not a rigid, but rather a situational and social, process that is subject to constant redefinition (cf. Brodbeck 2006, p. 5). For this reason, this work focuses on the limitation of the emergence of something new and not on the evaluation associated with it, thus excluding the limitation of this construct.

[1] To be specific, experience is not exactly meant, but ascribed. As a matter of fact, to experience something as a "cause", this means that associated perceptions are connected by the individual. Therefore, this does not mean cause in the first place, but the expectation of a relation, which could lead to the ascription "cause" to the antedate issue. However, for the differentiation made in this paragraph "to have experienced" is an adequate expression, for it should give the person reading the abstract idea of the concept.

At the same time, another aspect of creativity is evident in this social process. Because humans are living beings "because of their operational unity as non-trivial systems in the sense of Heinz von Foerster" (cf. Hejl 1992, p. 117), they cannot be trivialised, but remain dependent on inner states. These internal states are made possible because the increase in connections between receptors and effectors cited in footnote 2 of Chapter 8 leads to a development in which they begin to interconnect, as a result of which special forms such as the human brain may have formed via internal neuronal networks (cf. Maturana and Varela 2015, pp. 164f., 169ff., 173ff.).

Its self-referentiality is expressed as an inner part of thought. In this, the previous experiences, which are subject to an individual evaluation, are related. This reference is partly genetically predisposed, as can be shown by the cataract studies. In these, those unused areas are available again for the original use after years in which they were not developed corresponding to the purpose of the interconnection. The phases of prioritisation of incoming signals thus still seem to be open to the originally genetically planned signals (cf. Chatterjee 2015, p. 372f.). However, changes can also take place through the diverse use and collection of experience, as can be shown by the use of neurohormones altered by events (cf. Roth and Strüber 2014, p. 272). Similarly, in simulations, even minor modulations of state configurations after a short period of accumulation show very different development trajectories (cf. Rösler 2011, p. 397). For this reason, these inner states are provoked by the "perturbations" (Maturana and Varela 2015, p. 27); the autopoiesis, however, leads to the viable integration of this into what had existed until then, whereby the nervous system shows in contrast "to the autopoietic organism a much higher degree of variability" (Wagner 2007, p. 87) and is thus able to carry out state changes on a larger scale. These state changes also change the inner states to a certain extent, whereby recursions of operations are formed, which in a certain behavior start from the "root-cause" (in German, "Ur-Sache"), whereat the organism, not the stimulus, is responsible for the (final) behaviour (cf. von Foerster 1992, pp. 71, 74).

> "If in this process further interactions do not lead to state changes of the system anymore, then we have the case of the constitution or construction of objectivity by the interacting individual." (Hejl 1992, p. 124)

An individual has thus constructed his supposed objectivity. Through interaction with other individuals, however, it becomes possible to change this supposed objectivity. Because the coercive character of the social context, however, which already possesses fixed definitions of reality and the ideas of correct behaviour inherent in them as constructs, demands a one-sided change of state in the sense of adaptation, the definition of reality prevailing in these constructs is difficult to change (cf. Hejl 1992, p. 125). From this social sphere, communication can develop through the expression of symbols, and, if this communication is able to refer to itself, language. An important aspect of this is that language does not turn out to be a "language" because of the self-referentiality in its reference area as "reality per se, but a socially constructed or defined reality." (cf. Hejl 1992, p. 126). Therefore, however, as language is limited to the socially constructed "reality," an objectification of the social systems can only take place outside of it, because every observer in the network of social systems is always part of this network, and thus of this language and thus part of this constructed network *reality* (cf. Hejl 1992, p. 130). One way of uncovering these prefabricated structures may be an "approach-pluralistic and process-polymorphic contextualisation technique with action-reflexive self-knowledge interest" — psychoanalytics (cf. Greiner 2007, p. 139).

What does this "social interaction" have to do with the results? Well, here beautiful analogies to the neuronal network itself reveal themselves, which could certainly be further thoughts in the system–theoretical understanding. The single individual behaves in the social structure, which Hejl calls a network, as a node from which communication and interaction, that is, perturbation, takes place. In the brain, this is the task of interneurons, which represent an accumulation of neurons, that is, in a conditional way parts of the subjective experience and constructs. The individual in society is the same, a part of the subjective experiences and constructs of the whole. Both the interneuron and the individual can be both inhibiting and exciting to connect to other parts of the network. Thus, a part of the neuronal activity seems to be realised again in the realisation of social interaction. Without wanting to span this thought further, however, it should only be noted here that the transfer of the possibilities for the creation of something new can be thought not only intrapersonally but also interpersonally and thus seen as part of the necessity of strangification derived from constructivism.

Furthermore, the reality constructed by a person shows that the frames of the same also represent their limits. Therefore, the extension of this is only possible by strangifying it, by a reflection set in relation to another constructed reality and thus producing similarity and difference. This expansion of the "reality" is only possible in a social system that has produced language, because here communication itself can be reflected as the object of communication, and thus the constructed "reality".

Social systems are more flexible in their system reality than self-sustaining systems because of the nonpermanent expression of all their component states (cf. Hejl 1992, p. 135). It is hardly possible for the body to renounce breathing, but in thinking this is imaginable, just as it would be imaginable what would happen if one did not breathe for a long time. The biological body is limited as a self-sustaining system, but the mental system is not. Nevertheless, the limitations of the constructs will limit the mind. Furthermore, social systems through communication have a broader access to the environment, namely, to the wholeness of the constructed environment of the system, than the purely biological system (cf. Hejl 1992, p. 135). From this it can be concluded that, because social systems are not biological, an escape from the constructs is possible through them.[2] Because the structure of neuronal structures in self-referentiality can create a decoupling from the pure environment in analogy to

[2] At this point it should be parenthesised that social systems are per se pluralistic and multilayered. Consequently, different social understandings with differently developed reality definitions are realised in social systems. These understandings can build different structures, which experience reflection through confrontation as well as an extension or objectivity; however, this does not contradict the mentioned difficulties in adoption or change of these constructs. Contrariwise, the extension of social boundaries can lighten these.

this, there is an operational unity of the same, and "the nervous system functions as a closed network of changes in the activity relations between its components" (cf. Maturana and Varela 2015, p. 180), it is obvious that the creation of new realises an aspect of this outbreak in neuronal effectiveness.

In this context, language should be examined a little more closely. The foregoing described construct of the word "because", which introduces a special cause–effect relationship, is one experienced from the environment: it has been socialised and learned through interaction with other people. This socialisation goes so far that advantages can be obtained by the use of this word. Thus, in an experiment, people who asked a person at a printer to be allowed to pass were more likely to be allowed to pass if they incorporated the word "because" into the question. This condition nearly always occurred when the amount of material to be printed by the questioner was smaller than that of the questioned person, regardless of whether the "because" was followed by a meaningful reason or only by the statement "because I have to make copies" (cf. Langer et al. 1978, p. 636f.). This construct seems to be so powerful that under trivial conditions it is able to prevent rationality by attributing causality to the word.

At the same time, however, it is an abstract construct. A word like "because" is not a real one, but one to be experienced purely from the interaction of the environment. Under this separation, elements could be seen as descriptive and correlative. Words such as "house" would be direct words which, although they have different expressions, can still exist and be experienced as objects. Words such as "because" would be indirect, which can only be derived through the connection with the interaction of direct words, be it through changes taking place or attribution resulting from social interaction. Especially the indirect words become interesting for the constructions, because they imply (meaning) attributions and thus might represent the basis of the cause–effect affinity of humans, which is expressed, for example, in the difficulty of breaking out of these for quantum-physical processes in thinking (cf. Zeilinger 2005, p. 82ff.). Thus, it becomes increasingly clear that "reality, as we experience it, is a human creation in the sense that our entire experience is a human version of the world in which we live" (Schmidt 1992, p. 163).

Under this aspect, language is not the exchange of information or knowledge, but a special form of mutually adapted action (cf. von Glasersfeld 1992, p. 36). The constructed "reality" expresses itself in the language and hence is changed or proven by the social interaction that "this would [experience an] increase [in] their viability", much like this "is gained through repetition and coordination with different sensory impressions" (von Glasersfeld 1992, p. 37). This distinction makes it possible to check a model by checking it against other models, that is, not against the *reality* but only in the confrontation with other models (cf. Schmidt 1992, p. 157). In this context, language is part and foundation of cognitive structures "that are used in the structure of the experience". If these can also be ascribed in another subject (another subject is always a construct of ascriptions from the point of view of a subject) and are viable there, then "they gain a validity that we can justifiably call 'objective'" (cf. von Glasersfeld 1992, p. 37).

This reasoning means that the concepts mentioned in the previous differentiations are also to be regarded as constructs, but seem to be realised by different individuals from different fields, although they are also to be regarded as constructs. With this in mind, parallels have already been worked out in Chapter 6.1.2 common/shared terms which, although not in the depth of techniques such as the "Experimental Trans-Contextualization" (ExTC) in the "Standardized Therapy School Dialogue" (TSD) (cf. Greiner 2012), but at least attempted to relate the concepts inherent in the different models to one another to work out the similarities. Furthermore, a differentiation was created that could be further pursued in the investigation. This stringency makes it possible to assume that the results in terms of the effectiveness of the separation and the associated categorisation of the approaches for creating new things are likely to be reliable, which is reflected in the attribution in the *micro world* concerns.[3]

From this point of view, the results of the study become more understandable. No matter according to which subjective criteria the subdivision takes place, a change in the parameters creating these leads to a change in neuronal activities.

With this introduction, which has covered the categorisation of the results as well as the aspects of the model of thinking, it is now possible to proceed further in the discussion of the results. The second point of discussion is the obvious generalisation of the new. It is precisely the previous paragraph that suggests that a banalisation of creation would take place anew.

As a first starting point, the discussion of conceptuality, which was led by creativity, should be recalled here, in which the provision of ingenious services was mentioned as an essential aspect of extraordinary creativity. In creativity research, people like to talk about benefits (cf. Andreasen 2005, p. 17) or about applicability (cf. Carson 2010, p. 5). In the course of this discussion it was stated that this distinction can experience exciting distortions in experiments for problem solving, if a problem to be solved is combined with the unusual, that is, not everyday, use of objects. Overcoming such problems in itself already has the attribution of creativity. Here it becomes more obvious what was already mentioned briefly in the first discussion point, namely, that the separation between ordinary and extraordinary creativity is like walking a tightrope.

If one leaves the concept of creativity relatively broad in creativity research, then this makes sense, because "creative achievements and products are to be discovered in all human actions" (Brodbeck 2006, p. 5).

As a result, creativity research has increasingly focused on the process as well as the creative situation, which consists of external products, emotions, perceptions, movement patterns, and thought patterns (Brodbeck 2006, p. 5). The terms listed here can also be found in the test results. By combining the various models to form the model of thinking, it is thus possible to draw parallels here and confirm them.

---

[3] A detailed reflection of these approaches of philosophy of science es conducted in Chapter 11.1 (Reflection on Results from the Perspective of Philosophy of Science), in which the necessity of differentiation of similarity and difference related to the latter as foundation of deeper insight in the constructive realism is approached. Additionally, the deeper separation in the reality of the "micro world" and its validity is mentioned there understandably.

In the course of process research one cannot explain creativity "as a process of meaning, of experiencing positive-causal, but it can be negatively *said what* prevented *creativity effectively from the outside*" (Brodbeck 2006, p. 6). In contrast to this view, positive approaches can now be derived from this model of thinking. The possibilities for change based on the intervention of neuronal processes show that changes occur. These changes are partly owed to the inner and partly to the outer parts of thought, just as the various approaches to crossing the threshold of consciousness of elements that were not yet effective in asserting themselves were shown.

One reason for this creation of this positive approach could be that the relevant attribution of value in creativity research was deliberately excluded to avoid the "restriction of the scoring" (Brodbeck 2006, p. 9).

It can be seen here that creativity research is also aware that openness in the definition of creativity is necessary to not restrict it. Nevertheless, different paths are possible in the process: these have also already been mentioned in this work, the spontaneous and the (thoroughly) considered path (cf. Carson 2010, p. 60).

From this point of view, the omission of evaluation on the question of how something new and creative can emerge is justified insofar as the basis, namely, the emergence of something new, can be determined and is not subject to the restriction of an anticipated attribution of evaluation to social areas. "To predict something new in the future, however, would mean that the new is no longer new" (Brodbeck 2006, p. 8). This conclusion also implies the anticipatory attribution of valuation, which is per se obsolete in this work here.

Through the positive approach, the model of thinking assumed that the aspects derived from it for modifying the ability to create something new would be increased in reliability by reference to previous experiments. In contradiction to this, Brodbeck sees: "In any case, a model of the brain can neither predict actions nor creative products, and thus does not meet the requirements of an objective science that provides valid behavioral predictions" (Brodbeck 2006, p. 7). This statement was made in relation to the observation of the experiment of Libet, which would attribute the action initiative to the unconscious and thus define consciousness as a pure runaway with a limited right of veto (cf. Deecke and Soekadar 2016): it contradicts the approach that creativity processes can run purely unconsciously. The importance of consciousness is thus seen as the basis. This aspect was followed in this work in the definition of the thought to the extent that a thought presents itself as a vector constellation of internal and external parts realising in consciousness. At the same time, it became relevant in the context of the study that the various aspects of attention can bring individual information, which had previously been in the background, to the fore, that is, carry over the awareness threshold. Because there are different approaches to attention, these can only be realised in combination or interaction with other aspects.

The seemingly banal generalisation of the change of aspects has thus been relativised by the discussion about creativity and the deliberately omitted evaluation, which now leads to the third discussion point: the question about the changeability of the aspects of the model of thinking as a basis for the creation of something new.

Because thoughts are defined by their internal and external parts, which are able to penetrate into consciousness, as the study has shown, the three aspects can be addressed.

On the one hand, it is possible to attach the external components in such a way that the stimuli resulting from the external stimuli are presented in a way that would not otherwise be the case. In particular, this was possible, for example, through a change in posture or the special attention paid to external parts such as breathing in meditation. Different results were achieved by changing attention aspects, as could be expressed in the change of the reference level. From these approaches, different emotional influences as well as changes of the memory traces were produced that could lead to variations in the subsequent states.

On the other hand, the internal part of thoughts is to be called. The change of the internal parts of the thought can take place by changing a shift of attention in the intensive introspection, but also by the use of substances influencing the neuronal interconnection activities, emotional influences, irritations, language and context changes, script violations, or the omission of control hierarchies. These aspects are related to LTP/LTD as well as to the sustainable integration through neurogenesis and enable the creation of more efficient networked areas, which can serve as a basis for the easier random networking of new things as well as the connection generated by analogue thinking of otherwise very distant areas.

The third part has already been sounded, the consciousness. Because the discussion on awareness per se was not conducted (for the reasons cited in footnote 3 of Chapter 7), the implications to be derived here are purely from the expression of attention. The different states and forms of attention, focus, absence, and defocused focus are different amplifiers of new creation depending on the situation, in that they have significance for the carrying of otherwise too weak memory contents beyond the threshold of consciousness. It is therefore not unimportant that these states are used correctly according to the situation to be able to produce something new.

As can now be seen from these three summaries of the prerequisites of the aspects of thinking, this work claims to make a contribution from the pure avoidance tendency to the creation tendency in creativity research. In this sense, Brodbeck is quoted for the last time: "*Creative thinking is thinking outside the box. This often does not take place as a* more intended *but as the* allowance *of ascending ideas or changes in perception*" (Brodbeck 2006, p. 8).

In this sense, the expected behaviour is also predicted, at least with an increasing probability and tendency. But probably more is not to be expected from a discipline dealing with nontrivial machines in the sense of Heinz von Foerster — unless you take the nontriviality to absurdity.

Thus, the research-guiding question was answered to the extent that, excluding the concept of usefulness or evaluation from the definition of creativity, a three-part answer consisting of sub-aspects could be found as to how something new can emerge from thinking. The sub-questions were clarified as a prerequisite in the methodology, in which the consolidation of the models as well as the aspects connecting them were found.

References

Books and Journals

Andreasen, N. C. (2005). *The creating brain. The neuroscience of genius*. New York/Washington, DC: Dana Press.

Carson, S. (2010). *Your creative brain*. San Francisco: Jossey-Bass, A Wiley Imprint.

Chatterjee, R. (2015). Out of the darkness. Defying expectations, cataract surgery in Indian children in Indian children is endowing them with vision-and shedding light on how the brain learns how to see. *Science, 350*(6259), 372–375.

Deecke, L., & Soekadar, S. R. (2016). Beyond the point of no return: last-minute changes in human motor performance. *Proceedings of the National Academy of Sciences, 113*(21), E2876.

Greiner, K. (2007). Ist die psychoanalytische Neurophorie begründet? Zur epistemologischen Achillesferse der Neuro-Psychoanalyse. *Psychotherapy Forum*, 15, 134–140.

Greiner, K. (2012). *Standardisierter Therapieschulendialog (TSD). Therapieschulen-interdisziplinäre Grundlagenforschung an der Sigmund Freud Privatuniversität*. Vienna, Sigmund-Freud-Privatuniversitätsverlag.

Guttmann, G. (2000a). Die Messbarkeit des Erlebens und der Irrtum mit dem Ich. In F. Stadler (Ed.), *Elemente moderner Wissenschaftstheorie*. (pp. 73–91). Vienna/New York: Springer.

Hejl, P. M. (1992). Konstruktion der sozialen Konstruktion – Grundlinien einer konstruktivistischen Sozialtheorie. In H. Gumin & H. Meier (Eds.), *Einführung in den Konstruktivismus* (15th ed., pp. 109–146). Munich: Piper Publishers.

Langer, E., Blank, A., & Chanowitz, B. (1978). The mindlessness of ostensibly thoughtful action: The role of "placebic" information in interpersonal interaction. *Journal of Personality and Social Psychology, 36*(6), 635–642.

Maturana, H. R., & Varela, F. J. (2015). *Der Baum der Erkenntnis* (6th ed.). Frankfurt am Main: Fischer Paperback Publisher.

Rösler, F. (2011). *Psychophysiologie der Kognition. Eine Einführung in die Kognitive Neurowissenschaft*. Heidelberg: Spectrum Academic Publishing House.

Roth, G., & Strüber, N. (2014). *Wie das Gehirn die Seele macht* (5th ed.). Stuttgart: Klett-Cotta Publisher.

Schmidt, S. J. (1992). Vom Text zum Literatursystem – Skizze einer konstruktivistischen (empirischen) Literaturwissenschaft. In H. Gumin & H. Meier (Eds.), *Einführung in den Konstruktivismus* (15th ed., pp. 147–166). Munich: Piper Publishers.

von Foerster, H. (1992). Entdecken oder Erfinden – Wie lässt sich Verstehen verstehen?. In H. Gumin & H. Meier (Eds.), *Einführung in den Konstruktivismus*. (15th ed., pp. 41–88). Munich: Piper Publishers.

von Glasersfeld, E. (1992). Konstruktion der Wirklichkeit und des Begriffs der Objektivität. In H. Gumin & H. Meier (Eds.), *Einführung in den Konstruktivismus*. (15th ed., pp. 9–40). Munich: Piper Publishers.

Wagner, M. (2007). *Wir sehen mit den Augen des Kollektivs!?* Bad Heilbrunn: Julius Klinkhardt.

Zeilinger, A. (2005). *Einsteins Schleier – Die neue Welt der Quantenphysik*. Munich: Wilhelm Goldmann Publishers.

Online Resources

Brodbeck, K. H. (2006). Neue Trends in der Kreativitätsforschung. In *Themenschwerpunkt Kreativität, Kunst und Humor*. BÖP. Psychologie in Österreich, pp. 246-253. http://www.khbrodbeck.homepage.t-online.de/trends.pdf. Accessed 2 June 2016.

Chapter 10
Limitation of Validity

Contents

In the course of reference with the experiments, the integration of the models into an integrative whole and the resulting aspects of creativity were examined. Because the experiments were retrospectively considered under different conditions, just as they were treated only under the aspect of the averaged or otherwise reduced, albeit statistically correct, data, the results show a limitation of the generalisation, which is expressed in terms of the meaningfulness in relation to individual persons.

In the course of the investigation, the model of thinking first experienced a theoretical examination of the aspects derived from theory, which made it possible to elaborate theoretical foundations for the emergence of something new. These aspects were subsequently considered in various experiments as an explanation for the connection between the results of the experiments and the experimental setup. As a consequence, there is a danger that the subjective glasses of the limited construction will be used. In addition, it should be mentioned here that the consideration of the understanding of the respective model moments and thus of the reference setting constructed by the author as well as integration can be cited as limitations. However, because the aspects determined can be found in different models as well as linguistically in implicit or explicit parts by different authors, it can either be a general blind spot of the disciplines, or, because it can lead from different approaches to similar results, a viable construction under these conditions. If this were the case, it would be interesting in the sense that, in the course of the investigation of the experiments, new aspects could be revealed on the basis of the model of thought, which in this sense, for whatever reason, were not previously expressed. This finding would suggest that not only was the self-confirmation of the model carried out, but also an additional explanatory approach in the examination of empirical data

K. Testor, *Cognitive Theoretical Foundations of Creativity*,
https://doi.org/10.1007/978-3-030-61736-3_10

(banks). This conclusion means that the restriction of the subjective glasses should already be described as reduced.

A special restriction is placed on the theoretical execution of TRACT as the moment of the emergence of something new. Because this happens by chance, apart from extreme examples such as a stroke or the aforementioned persons with a sudden savant gift, it is hardly possible to make a reference when a sudden change occurs. As this is also the case on a very small scale, this consideration can only remain purely theoretical. More in-depth theoretical considerations as well as models that take this circumstance into account would still be conceivable but could hardly leave the theoretical level either. This circumstance results not only from the lack of accuracy in the resolution of the technical aids, but also from the fact that each brain is unique in its structure (cf. Finn et al. 2015, abstract). This uniqueness is both genetically predisposed and dependent on experience and behaviour. Similar to other chaotic, that is, multifactorial systems, the influence variables are so large in number that a prediction cannot be made any more than the exact propagation and relationship of the elements to each other.

In this sense, it should be pointed out once again that the question of the origin of thought cannot be answered either, but that it was only regarded as the basis of the processes in its biological expression of neuronal activity and is thus attempted to be understood under the restrictions derived from this system. It is not the brain that thinks. At least in part, thinking makes use of biological realisation in neuronal processes. These processes have come into focus in the attempt to understand as part of this work, but cannot explain, but only condense, the constructs used by this approach. In this context the discussion about consciousness or the body–soul dualism was deliberately omitted, because this would lead to the need for pursuing completely different traces of thought. As a prerequisite it was only assumed that there is a consciousness and that this can express itself by focusing in the attention, just as there are conscious and preconscious processes.

An additional limitation is the choice of models, as this is only one part of many different options. Theories such as the Theory of Mind (TOM), executive-process/interactive control (EPIC), capacity-constrained collaborative activation-based production system (3CAPS), or Soar, as well as the variants used for integration in the model of thought, are recognised in the discipline as having a right to exist for certain aspects. Whether or to what extent the aspects of thinking mentioned here would also be realised accordingly in these models can only be assumed in the sense of the constructivist discussion of the constructs grown from the scientific cultural tradition in the discussion of the results. In principle, however, the integration of several models should reduce the limitation of this model in the internal general validity of the discipline, especially because the aspects of working memory, declarative and procedural memory contents, memory traces, hierarchical control, and emotional and limbic systems are comprehensively depicted.

And perhaps the following applies to this: "that one should not invent things, sizes or entities for which there is no need" (Zeilinger 2005, p. 151f.). Occam's razor serves to prevent unnecessary aspects from being introduced into the process of finding theories and theses that would complicate it excessively or make it impossible to grasp.

Finally, it can still be said that applicability and experimental verifiability are necessary parameters for the empirical sciences, which is why an approach as yet not verifiable such as chance should be excluded. However, what would science be without thinking outside of boundaries?

References

Books and Journals

Finn, E. S., Shen, X., Scheinost, D., Rosenberg, M. D., Huang, J., Chun, M. M., ... & Constable, R. T. (2015). Functional connectome fingerprinting: identifying individuals using patterns of brain connectivity. *Nature Neuroscience, 18*(11), 1664–1671.

Zeilinger, A. (2005). *Einsteins Schleier – Die neue Welt der Quantenphysik.* Munich: Wilhelm Goldmann Publishers.

Chapter 11
Disciplinary Reintegration

Contents

Throughout these investigations, approaches were determined from the model of thinking as to how something new can emerge. But how can these ideas be of significance for the individual disciplines in the future?

As an introduction, it should be mentioned here that this is a theoretical model, which provides a fundamental explanation for the emergence of something new via various approaches, by changing the aspects of thinking that take place within the framework of neuronal processes and have been integrated up to that point. This approach makes it possible to make derivations for the individual persons, but also to make derivations for the interaction of persons, as far as it is questioned to what extent this interaction leads to something new arising.

11.1 Reflection on Results from the Perspective of Philosophy of Science

From the understanding of constructive realism, an essential aspect of science is interpretation, otherwise it would only be technology (cf. Wallner 2002, p. 139). In this context, it is necessary at this point to reflect on the conditions underlying this work, which should at least be outlined in a short essay, so that the conditions resulting from the "micro world" experienced limitation, more unconscious elements that

K. Testor, *Cognitive Theoretical Foundations of Creativity*,
https://doi.org/10.1007/978-3-030-61736-3_11

are always already loaded with theory as data (cf. Wallner 2002, p. 201) become rudimentarily conscious and can thus simultaneously limit the reconnection in science, but also allow it to gain quality in precision (cf. Wallner 2002, p. 249). Sciences have their own blind spots, which can be understood as their at that time to a "certain degree of precision and logical form" (Wallner 2002, p. 249) existing "blindness". If sciences/disciplines therefore became aware of these blind spots, they would gain the opportunity to partly overcome their own specialization caused "blindness", so that they could specify the limitations of the meaning of the results and thus better delimit their validity.

Why this reflection takes place in the disciplinary reconnection and not in the results discussion or the limitation of validity is briefly explained at this point. The reflection of the results should take place within a more general framework which, for the author, lies in the fundamentally necessary diversity of approaches and working methods in the natural sciences and humanities and cannot be found solely in the interpretation of the results of this work. Because these are general approaches, they should also be mentioned at the point at which they are related to other disciplines and at the same time show to what extent the natural scientists interested in the classifiability of their own products (cf. Wallner 2002, p. 193) can benefit from the strangification in the sense of a constructively realistic solution.

In this work it has already been pointed out several times that the *perception* of the environment in the actual sense cannot be connected with truth. Various experiments as well as insights into the functioning of nervous systems have shown that there are differences between a physical and a neuronally produced "reality". Here, the understanding of the nervous system expresses an already prepared realisation of the structural coupling, that is, a "history of reciprocal structural changes" (Maturana and Varela 2015, p. 85). This statement means that the nervous system, as a closed system, is to be understood as the "interaction area of an organism extended", which does not make itself up but is a "result of the changes in state that those perturbations trigger in it" (Maturana and Varela 2015, p. 156). Because a nervous system is in the course of a self-circuit created by interneurons, in addition to the signals varying from the inputs, further influences on the "perception" are created internally (cf. Maturana and Varela 2015, p. 178). At the same time, the nervous system is conceived in the sense of an operational unity, so that it works "as a closed network of changes in the activity relations between its components" (Maturana and Varela 2015, p. 180). This process enriches the autonomous nature of a living being by extending the otherwise existing operational unity of the same by an extension of the possibilities for action emanating from a nervous system. This finding demonstrates "that the nervous system does not absorb information, but produces information" (Wallner 2002, p. 196).

This work, as well as its approach, represents insights obtained from data, thus from empiricism. Now that the data are unambiguous, these could be regarded as generally valid. At the same time, however, the non-truthfulness of perception initiated by Maturana and Varela can be combined with scientific–theoretical trains of thought that make a relativisation of these clear.

At this point, here is the separation, understood by constructive realism, between "realität" and "wirklichkeit" explained.[1] The *realität* hence is the "world of our knowledge" and as "wirklichkeit" that world is understood, "with which we live, the world of our life's accomplishments in the broadest sense of the word" (cf. Wallner 2002, p. 141). Between these two is the lifeworld. This is also a constructed one, but at the same time it is taken as if it were a given (cf. Wallner 2002, p. 206): the "lifeworld is a system of beliefs and rules" (Wallner 2002, p. 208).

How is the living world created and why do we take it for granted? The basis of the structural coupling described by Maturana and Varela is interesting. This term expresses that when two organisms interact with each other, one of the two is to be seen as perturbation, that is, as disturbance without its negatively evaluating connotation (cf. Maturana and Varela 2015, p. 27, Note of the translators) of the other. Because a perturbation of organism A in organism B results in a change of state, which would have an effect on 'A' again in a retrospective process, where this again leads to the change, a cycle of changes is initiated. This cycle continues until some kind of equilibrium is reached, that is, a structural coupling, which includes partially the expression of the history of state changes. A similar point has already been made in this paper when reference was made to the recursion operations performed by von Foerster. These points showed that starting from a "root-cause" in the course of these loops a development of behaviour can be explained, but not from the cause itself, but from the organism. In the course of the recursions, it was shown that different initial values of a specific operation led to similar, to almost identical, results (cf. von Foerster 1992, p. 74f.).

This structural coupling is still comprehensible for simple operations or noncomplex systems, but it also takes place in more complex areas. A special feature is communication, which can be realised in a variety of ways, such as visually, acoustically, and olfactorily. Communication results in an interaction between organisms that is no longer based on direct action, but on an ascribed one, for understanding the construction of *wirklichkeit* and thus a fundamental basis for the world in which we live, as Wallner, referring to Wittgenstein, asks about the production of signs. Namely: What are signs and how do we know what they mean (cf. Wallner 2002, p. 145f.)? In this sense, signs are only a special realisation of communication to which a certain attribution of valuation or meaning is attributed; this is arbitrary but has grown over time. The starting point was coincidental, the realisation logical in the sense of a structural coupling. If this would not take place, no communication

[1] Wallner himself stated that there are "three key notions in constructive realism, namely wirklichkeit, realität and lebenswelt" (Wallner and Jandl 2006. p. 54). He further concluded that lebenswelt could be translated into English as lifeworld, but the other two do not have a distinguishable differentiation in English, which is why he himself did not translate but used the German words after an explanation. Until this point this separation was not necessary, but for the deeper understanding of the following implications the author uses the words that Wallner used himself. A deeper understanding can be found in the differentiation given in the work quoted above.

would be able to take place either, because this would cause a change as a perturbation in a counterpart.[2]

In the field of socially trained communication systems there is a special case, that of language, able to make itself the object of communication (cf. Hejl 1992, p. 126). Something exciting is happening for the theory of science. This recursive communication can be related to itself through recursiveness, and thus be extended beyond the actually limited area of information. The communal attribution of information requires a structural coupling expressed in socialisation or education. The socially "generated reality definitions and constructs" will be passed on by "*the* coercive *character of the social*" based on these constructs "aligned ideas of 'right' behavior" in "the one-sided change of state[:] adaptation" (Hejl 1992, p. 125) and thus handed down over time. Language as communication is thus a grown definition, and construction of reality and encompasses not only verbal but all other forms such as gestures. Now it should have become understandable why the world of life is constructed but at the same time is assumed to be given.

According to Einstein, the scientific world is a refinement of everyday beliefs. Wallner understands everyday beliefs as the lifeworld, the world in which people live (cf. Wallner 2002, p. 209). At the same time, nature does not provide knowledge, but rather its administration. "Knowledge requires nature management" (Wallner 2002, p. 161). Through science, this means that an administration of nature based on the selection of method and experiment-dependent construction of "reality" must take place. Science takes data carriers, that is, a certain amount of information, and a typesetting system that uses this information to tell us what happens when a certain event occurs, and thus constructs an artificial world in a specialist language, a "micro world" (cf. Wallner 2002, p. 211). Opposite the *wirklichkeit* the information obtained in this way has certain, and at the same time limited, properties (cf. Wallner 2002, p. 212). "Knowledge does not mean knowledge of nature, but how to structure nature" (Wallner 2002, p. 216).

This understanding has consequences. The *wirklichkeit* consists of an immense amount of data and information. It also has (relatively) unrestricted properties. The "realität" on the other hand is a world artificially created by us (cf. Wallner 2002, p. 211). In the language created by specialists, the starting point as well as the sentence systems based on it, the rules, are "theoretical buildings that are logically congruent in themselves". Therefore, the scientific experiences gained within this framework are considered to be true (cf. Wallner 2002, p. 204). However, the respective starting point is arbitrary; this arbitrariness is resolved by the formal application of rules (cf. Wallner 2002, p. 171). At the same time, something irritating occurs in

---

[2] Communication in this sense is confronted not only as social yet, which is a special segment to be examined soon, but in general. As an example, it also includes signal colours in animals or specially visible elements as a warning measure in poisonous plants. This kind of communication needs the ability to create a reasoning interpretation. Consequently, the poisoning in a consuming organism needs to rapidly lead to symptoms of poisoning or at least illness, which can be perceived by the organism itself as well as others and therefore initiate a learning process. This idea also shows how diverse reality can be.

relation to the consideration of a specific complex of themes. If different *micro worlds* examine a complex of topics, trivial information remains the same. The important ones, however, can change (cf. Wallner 2002, p. 205), because the life-world in its change also can and will have an effect on the micro worlds. The change in important areas is related to the change in formal rules. Because this changeability takes place within a culturally grown framework, because science as a subsystem of society is to be understood as a self-contained cultural framework, science is a cultural phenomenon and can only be understood within the respective cultural frameworks (cf. Wallner 2002, p. 197). In further consequence is the changing of this constructive stringing together of constructs in its cultural framework coherent, thus in its *micro worlds* it is considered true. The return of these micro worlds into the *wirklichkeit* sets the "constructivist misunderstanding" (cf. Wallner 2002, p. 136), because there is no comprehensible universally binding "wirklichkeit" that can be grasped.

Constructionism as a term explaining this misunderstanding in more detail is in the demarcation to constructivism purely concerning the object level, that is "a constructionist conception of the constitution of the world at the object level, which has no claim to explanation" (Wallner 2002, p. 134). Understanding this now circularly leads back to this work. Constructive realism is the meta-reflection created by speaking about language at the meta-level and by the suspension of legitimation (cf. Wallner 2002, p. 137). In this sense, there is no talk about the result of the scientific processes and their reality content, which, as should be apparent from the previous argumentation, is determined by the traditional and at the same time limiting life-world as well as the "realität" cannot be grasped, but the given is questioned and attempts to gain information about its structure by placing it in a new context (cf. Wallner 2002, p. 220f.). This act enables external criticism of an entire system, but not the detailed identification of errors (cf. Wallner 2002, p. 220). "The functioning, the correctness, the adequacy of the propositional systems is discussed by science and not by the philosophy of science" (Wallner 2002, p. 219).

This distinction is relevant in the term constructionism, because even within the micro worlds interpretations happen; otherwise, as mentioned at the beginning, they would only be technology and not science. However, these interpretations are already subject to a construct-based selection of rules that have grown over time. It is thus a micro-worldly culture that expresses itself as a structural coupling realised in this and thus creates a limitation. Because it is not possible to reflect within the system, methods such as strangification become necessary to make these limitations visible. Only through such an extension can an increased quality of knowledge and thus a somewhat broader validity be achieved. If one does not reflect on the structure, but purely on the objects, one does not leave the object level and thus also not constructionism.

At this point, Wallner's legitimation of the natural sciences in its micro-worlds as sciences are explained in brief, that is, interpreting them as scientific results. To be able to convey basic material, interpretation is needed, otherwise only basic material would be presented unchanged. An interpretation, however, is at the same time a change in the basic material and thus a reference to another. Thus, the interpretation in the broader sense corresponds to strangification (cf. Wallner 2002, p. 233). Although a pure application of different methods without further interpretation no longer leads to insight, only to mechanisation, the natural sciences already carry out strangification and thus interpretation in their methods. This is the attachment at the object level described in the foregoing paragraph, that is, constructionism, if this is not directed at the question of the type of structuring.

If now natural science in the interpretation, thus the strangification "is aware of his strangifying activity, one can begin to filter out to what extent strangification has been applied" (Wallner 2002, p. 233). In this application of the method of constructive realism, it becomes possible to work out and declare those factors of strangification and thus achieve an increase in the potential for knowledge. "In this way we approach the question of what of it is wirklichkeit and what is mere fiction " (loc. cit.). Wallner speaks of the *wirklichkeit* in the sense that it is "with whom we live, that is, everything that is there before our reflection, while realität is what we construct and what we reflect" (Wallner 2002, p. 234). He also says that it is "beyond human influence" (Wallner 2002, p. 233). "Wirklichkeit cannot serve as a category-forming theoretical justification, because the concepts of theory and category already belong to a specific understanding of science constructed by humans" (Wallner 2002, p. 234).

If one now consistently continues to think that, in constructive realism, the seemingly simple translation of scientific results from specialist language into everyday language, that is, into the lifeworld, is thus a huge reservoir for *strangification* (cf. Wallner 2002, p. 224), then it is above all about the deliberate translation. Similar to the translation of a term from one language into another, this reveals a certain amount of uncertainty (cf. Wallner 2002, p. 225). In the normal interdisciplinary discourse, such blurriness may be allowed for "pragmatic reasons" to stand "as they are". This is, however, a case of remnant rationality, an important institution against irrationality (cf. Wallner 2002, p. 149). It "prevails wherever there are structured relationships that cannot be discussed at the moment" (Wallner 2002, p. 148). But if these uncertainties are identified, a reflective gain in knowledge can result and insight into the underlying world can be achieved (cf. Wallner 2002, p. 225). Then, strangification is applied. Then the constructionist level is lifted to the constructivist level, because the reflection in the blur can leave the object relationship and the structure can be recognised.

What does this mean for this work? On the one hand, the sciences used, with the exception of psychotherapy, which occupies a special position in this sense, can be classified as natural sciences. Thus, they create a coherent theory building in their respective *micro worlds* through the sentence systems handed down over the years. At the same time, they create true statements in relation to the data sets derived from them. The grown concepts and understandings of the same are, as already hinted at in the results discussion, constructed separations. As such, they also have a tradition in the respective discipline. Whether these concepts are fiction or reality cannot be said through the purely interdisciplinary connection, as it is not clear whether and to what extent they have perturbed each other. The establishability of the terms shown as viable in this paper thus only expresses that the laws existing in the respective disciplines have been met, but not whether an increase in the potential for knowledge would be possible. By not elaborating the blurs, the object level is not left and therefore no real "strangification" is performed.

A brief example is the concept of memory content. If with this one a "strangification" would be carried out, for example, in the translation into everyday language, then the following dialogue could take place in such a short process. Memory contents are memories. – But memories are only there when you talk about them. – Well, there are also things that are stored in your memory that you do not tell at the moment. – Yeah, but then how do I know they are there? – It can be tested in experiments by querying them. – No, theoretically yes, but that is not possible, because then you would need as much time as you have lived until then. Moreover, telling stories would create new memories that one could not reproduce. That is not the point. But in theory, it is. – I've heard that some things are forgotten. – That can be, if one has certain disturbances in the network, that one loses something, and that then can no longer remember. – So then not everything can be remembered? – Actually not, but what is still stored in the brain is. – How is that actually stored? …

In this short symbolic dialogue, a circular journey was begun through mutual communication, which continues to move in the direction of a "structural coupling," that is, an understanding of the terms existing between these persons, in which the blurring becomes apparent at the same time. Through such a modus operandi it is possible to create knowledge. The sole different, methodical approach to a concept would not yet make it possible to leave one's own micro world and thus not yet condition the gain of knowledge about structuring made possible by reflection. Also, the connection of two disciplines on the pure object level would be to be attributed thereby only to the constructionism, because no level change takes place. The second level of constructive realism requires the suspension of legitimation, interpretation, openness, and diversity (cf. Wallner 2002, pp. 218, 220, 221, 223), and thus only a specific interpretation of data without extended reflection on the preconditions.

The suspension of the legitimation is carried out in such a way that the "instrumental basis of science as a special case" is seen as one which "should lie outside the philosophical judgment": this makes "realität" binding (Wallner 2002, p. 218).

Another important basis for this understanding is that experience cannot be standardised, because one cannot say when it is correct and when it is not, not even the scientific one: this results from the arbitrariness of the starting point and the subsequent dependence on the questions asked and the theories included (cf. Wallner 2002, p. 173). In addition, universals cannot be formulated, because a formulation of these would always at the same time mean a restriction and thus destroy a universal claim (cf. Wallner 2002, p. 167).

It has already been stated that the *wirklichkeit* is unmanageable in its diversity of data and information (cf. Wallner 2002, p. 211). Sciences take individual aspects from this diversity and investigate them. Thus, they limit the world as such and create a construction which is at the same time a limitation for understanding. Understanding this limitation is the gain in knowledge that wants to be created by constructive realism. "The scientific world is the lifeworld, made visible" (Wallner

2002, p. 253), and that is exactly what this work does in the micro world of the neurocognitive sciences, "realität" related to the question of how something new can emerge.

This point now leads to the question of which disciplines benefit in which way from this overview.

11.2 Reconnection

In this chapter, the recurrent connection to the disciplines named in Chap. 2, Disciplinary Connection, now takes place. Because the reflection on philosophy of science led to a deepening limitation of the work on the "micro world" of the neurocognitive sciences, the rebound in the individual disciplines with their "micro world" tradition of understanding, which normed and administrated specific aspects of the "wirklichkeit" in "realität" based on the empirical interpretation of data.

11.2.1 Medicine

An interesting approach for medicine is the thought experiment of random discharge, as when a neuron spontaneously discharges itself. The question posed in this context about the universality of the security of supply of the resumption of ions and the associated provision of energy could provide a basis for in-depth knowledge that could relativise the connection between ATPase disorders and cell death (cf. Yu et al. 2003, p. 2009). Furthermore, it would be conceivable in this context to gain insights into the relationship between glucose deprivation and stress, which could then be investigated.

In this context, this model provides an understanding of stress behaviour and the limited flexibility it provides, which is a basis for in-depth medical studies.

11.2.2 Cognitive Psychology

Through this model of thinking, an approach was created that not only is concerned with the networking of memory contents and their reproduction, but also, through their atypical activation, produces something new from itself. In this sense, this model brings an extension of the reference framework for experiments and their verifiability.

Furthermore, the understanding of memory contents has been expanded. These contents have been continued not only as chunks but as superchunks. Thus, the cascade structure, which results contextually in crosslinking, was explicitly named. In their connection, memory contents cannot be understood in isolation from context and level reference when they are realised. Although the omission of these in an intensive introspection allows the uncoupling of the external part of the thought, the internal still has an effect.

By working out the external and internal parts of thoughts as well as the necessity of crossing the threshold of consciousness to be reached for a thought, a definition was found that also represents a benefit for this discipline, if it is recognised that the receptors are not only directed outwards alone, but also explicate a special form of outwards within the body. This realisation makes connections such as the plurality of intestinal flora and the quantity of neurogenesis in the hippocampus (cf. Wolf et al. 2016) more understandable.

The isa-pointer of the chunks was also explained in more detail by the model. These chunks present themselves in more diverse ways and can carry emotional colourings as well as aspects of motor skills or posture. At the same time, these aspects also express changes in the relationships between applied production rules.

By incorporating the subdivision of the creative process into the part that produces and integrates the new, an approach was also indicated that can have an impact on teaching and learning research, as is especially true when considering the meditation experiment of Ding and colleagues. In this, the insight was brought to light that revising one's own unsuccessful or less successful approaches in the course of a nonevaluative openness leads to similar effects as the self-recognition in an *Aha!* experience (cf. Ding et al. 2015). In connection with the model of thinking, further progress could be made in didactics, in which the approaches could offer a possibility of generating a creative atmosphere.

A last part interesting for cognitive psychology should be mentioned here. The knowledge of efficient networks derived from this work has led to parallels between intelligence and expertise. These assumptions are based on similar underlying assumptions, which have resulted in higher efficiency. Such a move would be expected to facilitate activation, which would be expressed in a faster or easier spread of activation. In this context, the study by Thatcher and colleagues is interesting. In an EEG experiment with test persons, they have shown that the values measured in an IQ test strongly correlate with the "Phase Lock Duration" and "Phase Shift Duration."[3] A longer shift duration and a lower stability duration were associated with a higher IQ (cf. Thatcher et al. 2008, p. 1647). Furthermore, they showed that the significance of the correlations was higher in the near range (Thatcher et al. 2008, p. 1649). The derivations from this, in connection with the efficiency hypothesis as well as in connection with the realisation that a very high level of creativity with a very high IQ, except among physicians, is likely to have a disruptive/interfering effect (cf. Andreasen 2005, p. 12f.), would be an interesting basis for coming one step closer to the understanding of intelligence and creativity if this model is applied.

[3] Phase Lock Duration is the duration of phase stability in EEG-records, and Phase Shift Duration is the duration of the phase shift in EEG-records. The phase shift refers to the wandering of an oscillation activity measured at one point to another one (cf. Thatcher et al. 2008, p. 1642f.).

11.2.3 Neurosciences

For the neurosciences, this model created a more holistic connection by integrating Anderson's models, the four levels of personality, and the model of hierarchical control. This new model can be used again in the recurrent connection of experimental use. Thus, a model is also offered here for experimental reference which is suitable for creating the basis for something innovative from the pure reproduction of knowledge content through the creation of something new.

Furthermore, it would be likely that the main features of this work would also be present in animals. In this context, it would be interesting to take a closer look at this insofar as it could be reflected on investigations into problem solving and thus create finer coordination of the process. This work would be interesting in connection with the protein GFE3, which acts on inhibitory synapses, as it could produce a reversible elimination of this protein in vivo (cf. Gross et al. 2016, abstract).

An important aspect might be the context dependency of cascade formation and the networking of memory contents, which should have an effect on the progressive "mapping" of the brain. This idea is used, for example, in the "Human Connectome Project" in which a more and more precise assignment of areas and functions should become possible (cf. Glasser et al. 2016, p. 1). From this understanding, studies are also carried out in which speech is made possible with the help of signals derived from the brain (cf. Herff et al. 2015, p. 9). For this kind of understanding, on the one hand, this model provides a good basis for reflection with regard to the "representation" of content. In addition, with the help of expansion through the creation of something new and the necessary openness of the system, a basis for contextuality can be created and thus an understanding of nonrestriction of memory content. This change could help neuroscientific thinking to move away from ideas of a machine and pure representational thought to a more quantum-physical system. If, in this context, the constructive character of thinking, that is, its associatively constructive representation of constructed social constructs of a lifeworld mediated by language (in the sense of the constructivist discussion in the previous chapter), could be integrated, new possibilities would open up for this context, which from the author's point of view are important for deeper progress.[4]

[4]At this point it should be stated that certainly a limited use of words can be derived from a vague recording of brain activity. However, the explanation for this is not that science figured out where this information is stored, but that the brain caused by the perturbation was brought through these into a state of structural coupling and consequently a viable use occurred. Consequently, it cannot be named as understanding, but technically functional. As a consequence, a reflection of the structures and understandings beneath becomes more important, especially as already stated because of the individual uniqueness of the brain. This realization is insofar of relevance, as this peculiarity in its diversity can only be averaged but not predicted. In Chap. 10, this was indicated referring to Finn et al. (2015).

11.2.4 Psychotherapy Sciences

The special status of the psychotherapeutic sciences, especially psychoanalysis, cited in the reflection of results from a philosophy of science point of view is based on the fact that they use a circular, that is, hermeneutic, approach in their application (cf. Wallner 2002, p. 251). The discussion in this respect as well as the attempt to connect to the natural sciences are not carried out here. However, one benefit for psychoanalysis as well as for the other therapy schools can be seen in that the critical philosophical reflection in this work should show that a different claim is to be made to the collection of data and their interpretation as a nature-administrating achievement than in the distinction emerging and reflecting from the system, which is about gaining knowledge about the possibilities of structuring nature. This second point should be taken up by psychotherapy sciences, because "the traditional notion of a universally binding truth" (Guttmann 2000a, p. 5) must define the thoughts of a human being as constructs. From this, it can be concluded that in therapy one can only look behind the structure of these constructs, but not at their binding nature.

As a consequence, this model of thinking can serve both as a context for strangification and as an approach, under the aspect of limited validity, in the reconnection to the lifeworld, to how practical "strangification" can be carried out in patients by changing the context in a conversation or by using various methods mentioned in this paper. This method makes it possible to work out the structure behind the constructs.

Another point that can be made here is that the "neuro-phoria" (cf. Greiner 2007) dedicated approaches can experience a critical reflection of their own possibilities, which, similar to the understanding of representation cited in the return to the neurosciences, should experience a gain in knowledge through constructive realism. In this sense the same shift away from the idea of representation would also be necessary here and with it the rejection of the possibility of universals.

11.2.5 Creativity Research

The model of thinking offers creativity research a broader approach to experimental design through the construction resulting from the integration of several models. In addition, the approaches explicated in this model can be a gain in creativity research for further consideration.

An important aspect here is the previously reduced definition of creativity around the moment of evaluation, which certainly makes sense for extraordinary creativity, but causes a strongly restricting limitation in understanding for ordinary creativity and thus for the emergence of something new. Thus, an approach was created that can go beyond the negative understanding of organisational frameworks for the creation of creativity, thus not only designates what disturbs creativity, but on the contrary, can offer constructive possibilities.

This constructive approach to explaining the creation of something new makes it possible to develop further creativity techniques and approaches from this, which can then also be applied in practice, while at the same time enabling a deeper understanding of the theoretical superstructure.

Finally, the gain in knowledge for process research that can be derived from these aspects should be mentioned, which can now be assigned to the various phases, that is, preparation, incubation, or the determined search for solutions to problems, and finally verification and integration (cf. Carson 2010, p. 62).

11.2.6 Philosophy of Science

For the philosophy of science, this offers a possibility of extending the potential for strangification by deriving creative techniques from the aspects of this model. This benefit lies in the fact that the approaches to creating something new hinted at in the aspects can do precisely that: generate a strangification. If a change of context takes place, then the recursions, if one allows and responds to them, begin to change in this context, leading to a deepening understanding and to recognising the underlying structure.

As "experimental hermeneutic thinking, research and analysis" stimulate "the ability to shape and express creatively," so "understanding meaning [functions] by strangifying meaning" (Greiner and Jandl 2015, p. 7), which are in the constructively realistic sense for this form of reflection alongside the already existing techniques such as the *Experimental Trans-Contextualization* (ExTC) in the "Standardized Therapy School Dialogue" (TSD) (cf. Greiner 2012), more "playful-creative analysis modes" were recently developed. The text puzzle methods, the image process methods, or the various media games can only be mentioned here as a fraction. In a similar way, the aspects can serve as inspiration to expand this "bizarrosophical" variants or to create others.

Even under the assumption that the truthfulness of the aspects of the model of thought is based on the micro world of the neuro- and cognitive sciences, they can at least serve as a starting point for such extensions. After all, the philosophical scientists of constructive realism know that the starting point and the rules are arbitrary, but the arbitrariness dissolves under their application in the following ways of thinking. "A way of thinking is one consistently selected from the multitude of possible ways of thinking" (Wallner 2002, p. 171).

11.2.7 Other

Thinking is the inner language that has grown from social interaction, and thus presupposes language, which in turn presupposes (perturbing) social interaction/communication, which is realised in a form corresponding to the living world.

Because language and thus thinking are to be seen as a basic characteristic of humans, derivations can be made for all areas that work with humankind itself. This derivation can serve as a means of constructivist reflection as well as through the connection without deepening reflection in the interdisciplinary dialogue on the object level, that is, constructionist.

This universal claim is valid to the extent that in all sciences the expansion of one's own *micro world* is valid so long as it is complying with one's own constructs and that these experience interpretation in the area, whereby an interpretation and insight can be achieved by changing the initial data (cf. Wallner 2002, p. 233).

References

Books and Journals

Andreasen, N. C. (2005). *The creating brain. The neuroscience of genius*. New York/Washington, DC: Dana Press.

Carson, S. (2010). *Your creative brain*. San Francisco: Jossey-Bass, A Wiley Imprint.

Ding, X., Tang, Y. Y., Cao, C., Deng, Y., Wang, Y., Xin, X., & Posner, M. I. (2015). Short-term meditation modulates brain activity of insight evoked with solution cue. *Social Cognitive and Affective Neuroscience, 10*, 43–49.

Greiner, K. (2007). Ist die psychoanalytische Neurophorie begründet? Zur epistemologischen Achillesferse der Neuro-Psychoanalyse. *Psychotherapy Forum*, 15, 134–140.

Greiner, K. (2012). *Standardisierter Therapieschulendialog (TSD). Therapieschulen-interdisziplinäre Grundlagenforschung an der Sigmund Freud Privatuniversität*. Vienna, Sigmund-Freud-Privatuniversitätsverlag.

Greiner, K., & Jandl, M. J. (2015). *Bizarrosophie – Radikalkreatives Forschen im Dienste der akademischen Psychotherapie*. (1st ed.). Nordhausen: Traugott Bautz.

Gross, G. G., Straub, C., Perez-Sanchez, J., Dempsey, W. P., Junge, J. A., Roberts, R. W., ... & Sabatini, B. L. (2016). An E3-ligase-based method for ablating inhibitory synapses. *Nature Methods, 13*, 673–678.

Guttmann, G. (2000a). Ludwig von Bertalanffys Systemtheorie und die „Wirklichkeit" psychologischer Konzepte. In K. Edlinger, W. Feigl, & G. Fleck (Eds.), *Systemtheoretische Perspektiven*. Frankfurt am Main: Peter Lang. Manuscript of the author.

Hejl, P. M. (1992). Konstruktion der sozialen Konstruktion – Grundlinien einer konstruktivistischen Sozialtheorie. In H. Gumin & H. Meier (Eds.), *Einführung in den Konstruktivismus* (15th ed., pp. 109–146). Munich: Piper Publishers.

Maturana, H. R., & Varela, F. J. (2015). *Der Baum der Erkenntnis* (6th ed.). Frankfurt am Main: Fischer Paperback Publisher.

Thatcher, R. W., North, D. M., & Biver, C. J. (2008). Intelligence and EEG phase reset: a two compartmental model of phase shift and lock. *Neuroimage, 42*, 1639–1653.

von Foerster, H. (1992). Entdecken oder Erfinden – Wie lässt sich Verstehen verstehen?. In H. Gumin & H. Meier (Eds.), *Einführung in den Konstruktivismus*. (15th ed., pp. 41–88). Munich: Piper Publishers.

Wallner, F. (2002). *Die Verwandlung der Wissenschaft*. Jandl, M.J. (Hg.). Hamburg: Dr. Kovač.

Wallner, F. G., & Jandl, M. J. (2006). The importance of constructive realism for the indigenous psychologies approach. In U. Kim, K.-S. Yang, & K. K. Hwang (Eds.), *Indigenous and cultural psychology* (pp. 49–72). New York: Springer.

Wolf, S. A., et al. (2016). Ly6Chi monocytes provide a link between antibiotic-induced changes in gut microbiota and adult hippocampal neurogenesis. *Cell Reports, 15*(9), 1945–1956.

Yu, S. P., Wang, X. Q., Xiao, A. Y., Sheline, C., Hyrc, K., Yang, A., ... & Goldberg, M. P. (2003). Apoptotic insults impair Na+, K+-ATPase activity as a mechanism of neuronal death mediated by concurrent ATP deficiency and oxidant stress. *Journal of Cell Science, 116*, 2099–2110.

Online Resources

Glasser, M. F., Coalson, T. S., Robinson, E. C., Hacker, C. D., Harwell, J., Yacoub, E., ... & Smith, S. M. (2016). A multi-modal parcellation of human cerebral cortex. *Nature, 536*(7615), 171-178. https://doi.org/10.1038/nature18933.

Herff, C., Heger, D., De Pesters, A., Telaar, D., Brunner, P., Schalk, G., & Schultz, T. (2015). Brain-to-text: decoding spoken phrases from phone representations in the brain. *Frontiers in Neuroscience, 9*, 217. https://doi.org/10.3389/fnins.2015.00217.

Chapter 12
Résumé

Contents

Based on the cognitive sciences and their models, a new model was integrated into the neurosciences on the basis of neuroanatomical principles, which differed from its predecessors. Taking basic principles into account, this model connects both the memory contents in their networking and their tendency to spread, referred to as memory traces. Furthermore, it connects the preparatory role, which comes from the limbic system via an integration, as well as the production rules working from these aspects, which form a special hierarchical control through the interconnection structure, with the neuronal modes of action.

This integration has led to approaches for changes in interconnection patterns, which, in relation to the definition of thought as a special vector constellation of the neuronal activity pattern generated by an external and internal part, which is perceived in consciousness, have been attached to its parts.

The reference of these aspects of the model of thinking developed from the models could be linked with various experiments and could theoretically be confirmed as an explanatory approach for the results of the experiments, as well as the resulting conclusions for the creation of something new.

This finding showed that chunks and production rules are to be understood as elements constructed from the environment that represent implicit and explicit content. Chunks are connected to each other through an isa-pointer (cf. Anderson 1996b, p. 355f.).

K. Testor, *Cognitive Theoretical Foundations of Creativity*,
https://doi.org/10.1007/978-3-030-61736-3_12

Here is one of the starting points that can be achieved both by changing the context and thus by changing the extent of memory traces (cf. Anderson 1996a, p. 178 in conjunction with Godden and Baddeley 1975, passim). The emotional colouring can also serve as a context, as was derived from Ekman and Davidson's experiment with body language. A change in these areas also leads to changes in the neuroanatomy, as LTP/LTD and integration through neurogenesis takes place (cf. Kahle and Frotscher 2005, p. 30; Silbernagl and Despopoulos 2003, pp. 336, 42, 10, 12 in connection with Roth and Strüber 2014, pp. 254–256; Shors 2009, passim).

For the creation of something new, attention is an important aspect that may be expressed in different ways depending on the progress of the process. Attention acts both as an amplifier and as a damper (cf. Luck et al. 1994, p. 898f.). For example, reduced attention is an advantage if it does not suppress optional areas (cf. Rösler 2011, pp. 91, 99f; Simons and Chabris 1999, passim) and can thus help other active areas rise from the background to the foreground. Increased attention is also advantageous if the focus on these weaker signals can help them to break through into consciousness, as can be seen from the example of the Stroop effect (cf. Anderson 1996a, p. 94ff.). As a further example of attention, it was shown that it can combine these two effects in a defocused focus, which was achieved, for example, through mindfulness techniques. The reference to the process phase results from the fact that in the preparation as well as in the verification concentration is more important than pure letting flow. In the incubation phase, this is vice versa.

A change of the hierarchical control leads to a reduction of the inhibition of the interneurons by lateral disinhibition, whereby more distant areas can connect with each other (cf. Carson et al. 2003, abstract). This connection also changes the spread of memory traces, as could be shown in the experiment with alcohol using the word association test. The changes in this area can therefore be caused by stimulants, which is not recommended, but also by training and practice (cf. Carson 2010, preface, p. X, and Carson 2010, p. 108; Ding et al. 2015 in connection with Tang et al. 2010, p. 15650f.).

An important point in the creation of something new is the probability of activation, which can also be found in the efficiency of the circuitry. This point was shown on the one hand by the thought experiment of theoretical random activation (TRACT) and on the other hand by the experiment for the integration of complete strangeness via analogies, which might relate previously known knowledge to something new. Therefore, the everyday observation that very successful people with high creative potential have attained a high level of skills and abilities in their fields is also confirmed in this area, because this can be a beneficial aspect for creating something new, although it is not a prerequisite.

The creation of new things can be explained by the wiring pattern of the brain. An important aspect of this work is that this approach is based on the historically grown culture of neuro- and cognitive sciences. Therefore, this was restricted to this object area in the sense of a scientific–theoretical reflection in general validity. This limitation is a constructionist one because, although various methods were used and related, no activity was carried out that reflected the underlying structure. Thus, the insight of this work is based on the *micro world* but can still be used as a basis for other constructionist and constructivist models.

Taking this point into account, it can be said that this work offers a good opportunity for a positive approach to process research in creativity research, as well as the neurosciences and the cognitive sciences, an extension of their approaches beyond purely reproductive thought processes.

References

Books and Journals

Anderson, J. R. (1996a). *Kognitive Psychologie. Eine Einführung.* (2nd ed.). Heidelberg: Spectrum Academic Publishing House Ltd.

Anderson, J. R. (1996b). ACT. A simple theory of complex cognition. *American Psychologist, 51*(4), 355–365.

Carson, S. (2010). *Your creative brain.* San Francisco: Jossey-Bass, A Wiley Imprint.

Carson, S. H., Peterson, J. B., & Higgins, D. M. (2003). Decreased latent inhibition is associated with increased creative achievement in high-functioning individuals. *Journal of Personality and Social Psychology, 85*(3), 499–506.

Ding, X., Tang, Y. Y., Cao, C., Deng, Y., Wang, Y., Xin, X., & Posner, M. I. (2015). Short-term meditation modulates brain activity of insight evoked with solution cue. *Social Cognitive and Affective Neuroscience, 10*, 43–49.

Godden, D. R., & Baddeley, A. D. (1975). *Context dependent memory in two natural environments: on land and underwater.* Stirling: University of Stirling. Department of Psychology. Revised manuscript.

Kahle, W. & Frotscher, M. (2005). *Nervensysteme und Sinnesorgane.* In: Taschenatlas Anatomie. 3rd band (9th revised edition). Stuttgart. Thieme Publishers.

Luck, S. J., Hillyard, S. A., Mouloua, M., Woldorff, M. G., Clark, V. P., & Hawkins, H. L. (1994). Effects of spatial cuing on luminance detectability: psychophysical and electrophysiological evidence for early selection. *Journal of Experimental Psychology: Human Perception and Performance, 20*(4), 887–904.

Rösler, F. (2011). *Psychophysiologie der Kognition. Eine Einführung in die Kognitive Neurowissenschaft.* Heidelberg: Spectrum Academic Publishing House.

Roth, G., & Strüber, N. (2014). *Wie das Gehirn die Seele macht* (5th ed.). Stuttgart: Klett-Cotta Publisher.

Shors, T. J. (2009). Saving new brain cells. *Scientific American, 300*(3), 47–54.

Silbernagl, S. & Despopoulos, A. (2003). *Taschenatlas der Physiologie* (6th corrected edition). Stuttgart: Thieme Publishers.

Simons, D. J., & Chabris, C. F. (1999). Gorillas in our midst: sustained inattentional blindness for dynamic events. *Perception, 28*(9), 1059–1074.

Tang, Y. Y., Lu, Q., Geng, X., Stein, E. A., Yang, Y., & Posner, M. I. (2010). Short-term meditation induces white matter changes in the anterior cingulate. *Proceedings of the National Academy of Sciences, 107*(35), 15649–15652.

Chapter 13
Outlook

Contents

The connection between thinking and creativity, that is, the new that arises from thinking, is a very exciting question. This work offers an approach based on constructs from the neurosciences and cognitive sciences that connects thinking with its biological prerequisite, the brain. However, the brain is not thinking, but only a necessary functional basis for this. If one thinks at this point of those people who are apparently able to live normally without a fully developed brain (cf. Feuillet et al. 2007, passim), these approaches are relativised just as it is the case by the savants who realize unimaginable talents, and this apparently from one day to the next, triggered solely by a thrust in the truest sense of the word (cf. Piore 2013, passim).

Stress is a term that is often used in current times. Because stress restricts cognitive performance (cf. Sliwinski et al. 2006), as does high-fat nutrition (cf. Jais et al. 2016) or multitasking (cf. Spitzer 2012, p. 229), the need for countermeasures or prevention will also increase in the future, whether for economic or therapeutic reasons. To be prepared for this time, this model could be advantageous in conjunction with findings on cognition-promoting factors such as playing an instrument (cf. Wan and Schlaug 2010), education, exercise, and healthy nutrition (cf. Spitzer 2012, p. 298). Through this connection, the conducive factors could be viewed from the aspect of creation of the new and thus form a new perspective.

Starting from the topic of stress, approaches for changing framework conditions are conceivable from this model, as they have already been demonstrated in the area of PTSS through the change of physical framework conditions (cf. Cook et al. 2016). Similarly, the well-being hormones that increase through the creation of

K. Testor, *Cognitive Theoretical Foundations of Creativity*,
https://doi.org/10.1007/978-3-030-61736-3_13

something new and the activation of the reward centre would be conceivable as an approach for countermeasures (cf. Ding et al. 2015).

In the context of diversity management, this model could provide interesting connections with group dynamic processes. Because good diversity management can achieve better performance than more homogeneous groups and this results from better mental flexibility (cf. Jans 2004), this process can be linked to this model.

In this context, one possibility is to be found in the connection with decision-making processes of executives at different management levels, whereby the focus would be on those areas that would allow for a corresponding room for manoeuvre, as is the case, for example, with strategic decisions. For these areas, this model can be used as a support both in the process and for investigations. It can also serve as an approach to explaining processes.

It was already suggested during the disciplinary reintegration that an interesting connection could be made between this model of thinking and the question of the efficiency of neuronal structures. Although this is not dependent on the model in the basic question and would certainly provide information about the connections between intelligence and expertise, in a connection with the IQ threshold based on these elements, a basic understanding of thinking can be deepened and connected with new insights in relation to creativity and in conjunction with the increased susceptibility of creative persons to mental illness.

Of course, new creative techniques can be developed using this model, and existing models, can be expanded. This result is to be expected to the extent that this area has no restriction to expect from the discipline as well as from the area of life. The "bizarrosophy", which can be understood as a playful analysis system in which creative techniques that seem to be bizarre are used to create an "strangification" and thus a gain in knowledge for experimental hermeneutics, can be created (cf. Greiner and Jandl 2015).

These new techniques can also be used to think about progress in teaching as well as in learning, because this model could be used to derive approaches for better integration of new knowledge content. How these findings could be integrated into the existing concepts or how they could be expanded is certainly an exciting question.

For the cognitive sciences, this model allows a change in the approach to understanding thinking. To get away from the processing of reproductive performances in experimental design, this model can be used to pursue the creation of something new via the individual aspects. Thus, from the field of constitution starting from memory, an extension is possible that could confront the question of how creativity, possibly even specific, from prepared situations can be expected.

The last point to be mentioned is the consciously omitted aspect of consciousness. As this question is very complex and much discussed in various disciplines, the question arises what consciousness is and how it expresses itself. In this work, consciousness was conceived as a necessary prerequisite for a thought without any further delimitation, and was even used in the definition of thought. It remains to be seen whether this model of thinking, which has been based on hierarchical control in relation to the enforcement of activity patterns in the background, can provide an approach to this topic. The author himself has only one assumption at this point, which would rather go in the direction of quantum-physical understanding. Nevertheless, it would be nice if, from the understanding expressed here in terms of areas, a part could be contributed to a development in this direction.

References

Books and Journals

Cook, I. A., Abrams, M., & Leuchter, A. F. (2016). Trigeminal nerve stimulation for comorbid posttraumatic stress disorder and major depressive disorder. *Neuromodulation, 19*(3), 299–305.

Ding, X., Tang, Y. Y., Cao, C., Deng, Y., Wang, Y., Xin, X., & Posner, M. I. (2015). Short-term meditation modulates brain activity of insight evoked with solution cue. *Social Cognitive and Affective Neuroscience, 10*, 43–49.

Feuillet, L., Dufour, H., & Pelletier, J. (2007). Brain of a white-collar worker. *The Lancet, 370*(9583), 262.

Greiner, K., & Jandl, M. J. (2015). *Bizarrosophie – Radikalkreatives Forschen im Dienste der akademischen Psychotherapie.* (1st ed.). Nordhausen: Traugott Bautz.

Jais, A., Solas, M., Backes, H., Chaurasia, B., Kleinridders, A., Theurich, S., ... & Alber, J. (2016). Myeloid-cell-derived VEGF maintains brain glucose uptake and limits cognitive impairment in obesity. *Cell, 165*(4), 882–895.

Jans, M. (2004). Empirische Effekte organisationsdemografischer Diversität in Organisationen – Ergebnisse und Erkenntnisse einer Metaanalyse. Essener Beiträge zur Personalforschung Nr.3. In *soFid Organisations- und Verwaltungsforschung 2006/2*. (pp. 9–43). GESIS-Leibniz-Institute for Social Sciences.

Sliwinski, M. J., Smyth, J. M., Hofer, S. M., & Stawski, R. S. (2006). Intraindividual coupling of daily stress and cognition. *Psychology and Aging, 21*(3), 545–557.

Spitzer, M. (2012). *Digitale Demenz.* Munich: Droemer Publishers.

Wan, C. Y., & Schlaug, G. (2010). Music making as a tool for promoting brain plasticity across the life span. *The Neuroscientist, 16*(5), 566–577.

Online Resources

Piore, A. (2013). When brain damage unlocks the genius within. *Popular Science, 2013.* https://www.popsci.com/science/article/2013-02/when-brain-damage-unlocks-genius-within/ Accessed 21 Nov 2020.

References

Books and Journals

Anderson, J. R. (1996a). *Kognitive Psychologie. Eine Einführung.* (2nd ed.). Heidelberg: Spectrum Academic Publishing House Ltd.

Anderson, J. R. (1996b). ACT. A simple theory of complex cognition. *American Psychologist, 51*(4), 355–365.

Anderson, J. R., Bothell, D., Byrne, M. D., Douglass, S., Lebiere, C., & Qin, Y. (2004). An integrated theory of the mind. *Psychological Review, 111*(4), 1036–1060.

Andreae, L. C., & Burrone, J. (2015). Spontaneous neurotransmitter release shapes dendritic arbors via long-range activation of NMDA receptors. *Cell Report, 10*(6), 873–882.

Andreasen, N. C. (2005). *The creating brain. The neuroscience of genius.* New York/Washington, DC: Dana Press.

Arns, M., De Ridder, S., Strehl, U., Breteler, M., & Coenen, A. (2009). Efficacy of neurofeedback treatment in ADHD: the effects on inattention, impulsivity and hyperactivity: a meta-analysis. *Clinical EEG and Neuroscience, 40*(3), 180–189.

Athanasopoulos, P., Bylund, E., Montero-Melis, G., Damjanovic, L., Schartner, A., Kibbe, A., ... & Thierry, G. (2015). Two languages, two minds: Flexible cognitive processing driven by language of operation. *Psychological Science, 26*(4), 518–526.

Baddeley, A. (2003). Working memory: looking back and looking forward. *Nature Reviews Neuroscience, 4*, 829–839.

Badre, D., & D'Esposito, M. (2009). Is the rostro-caudal axis of the frontal lobe hierarchical? *Nature Reviews Neuroscience, 10*, 659–669.

Bouhuys, A. L., Bloem, G. M., & Groothius, T. G. G. (1995). Induction of depressed and elated mood by music influences the perception of facial emotional expression in healthy subjects. *Journal of Affective Disorders, 33*(4), 215–226.

Bower, G. H. (1981). Mood and memory. *American Psychologist, 36*(2), 129–148.

Brozinsky-Schwabe, E. (2011). *Interkulturelle Kommunikation.* Wiesbaden: Publishing House for Social Sciences.

Burton, H., Snyder, A. Z., Conturo, T. E., Akbudak, E., Ollinger, J. M., & Raichle, M. E. (2002). Adaptive changes in early and late blind: a fMRI study of Braille reading. *Journal of Neuroscience, 87*(1), 589–607.

Cahn, R. B., & Polich, J. (2006). Meditation states and traits: EEG, ERG, and neuroimaging studies. *Psychological Bulletin, 132*(2), 180–211.

Carletti, B., & Rossi, F. (2008). Neurogenesis in the cerebellum. *The Neuroscientist, 14*(1), 91–100.

Carson, S. (2010). *Your creative brain.* San Francisco: Jossey-Bass, A Wiley Imprint.

K. Testor, *Cognitive Theoretical Foundations of Creativity*,
https://doi.org/10.1007/978-3-030-61736-3

Carson, S. H., Peterson, J. B., & Higgins, D. M. (2003). Decreased latent inhibition is associated with increased creative achievement in high-functioning individuals. *Journal of Personality and Social Psychology, 85*(3), 499–506.

Chase, W. G., & Simons, H. A. (1973). The mind's eye in chess. *Visual Information Processing, 4*, 215–281.

Chatterjee, R. (2015). Out of the darkness. Defying expectations, cataract surgery in Indian children in Indian children is endowing them with vision-and shedding light on how the brain learns how to see. *Science, 350*(6259), 372–375.

Chepenik, L. G., Cornew, L. A., & Farah, M. J. (2007). The influence of sad mood on cognition. *Emotion, 7*(4), 802–811.

Cook, I. A., Abrams, M., & Leuchter, A. F. (2016). Trigeminal nerve stimulation for comorbid posttraumatic stress disorder and major depressive disorder. *Neuromodulation, 19*(3), 299–305.

Cryan, J. F., Bravo, J. A., Forsythe, P., Chew, M. V., Escaravage, E., Savignac, H. M., ... & Dinan, T. G. (2011). Ingestion of Lactobacillus strain regulates emotional behavior and central GABA receptor expression in a mouse via the vagus nerve. *Proceedings of the National Academy of Sciences, 108*(38), 16050–16055.

Davies, M. (2003). The role of $GABA_A$ receptors in mediating the effects of alcohol in the central nervous system. *Journal of Psychiatry & Neuroscience, 28*(4), 63–74.

De Bellis, M. D., Clark, D. B., Beers, S. R., Soloff, P. H., Boring, A. M., Hall, J., ... & Keshavan, M. S. (2000). Hippocampal volume in adolescent-onset alcohol use disorders. *American Journal of Psychiatry, 157*(5), 737–744.

Deecke, L., & Soekadar, S. R. (2016). Beyond the point of no return: last-minute changes in human motor performance. *Proceedings of the National Academy of Sciences, 113*(21), E2876.

Dejana, E., & Betsholtz, C. (2016). Oligodendrocytes follow blood vessel trails in the brain. *Science, 351*(6271), 341–342.

Ding, X., Tang, Y. Y., Cao, C., Deng, Y., Wang, Y., Xin, X., & Posner, M. I. (2015). Short-term meditation modulates brain activity of insight evoked with solution cue. *Social Cognitive and Affective Neuroscience, 10*, 43–49.

Ekman, P., & Davidson, R. J. (1993). Voluntary smiling changes regional brain activity. *Psychological Science, 4*, 342–345.

Eysenck, M. W. (2009). *Fundamentals of psychology*. East Sussex-New York: Psychology Press.

Fademrecht, L., Bülthoff, I., & de la Rosa, S. (2016). Action recognition in the visual periphery. *Journal of Vision, 16*(3), 33.

Ferris, L. T., Williams, J. S., & Shen, C. L. (2007). The effect of acute exercise on serum brain-derived neurotrophic factor levels and cognitive function. *Med Sci Sports Exerc, 39*(4), 728–734.

Feuillet, L., Dufour, H., & Pelletier, J. (2007). Brain of a white-collar worker. *The Lancet, 370*(9583), 262.

Finn, E. S., Shen, X., Scheinost, D., Rosenberg, M. D., Huang, J., Chun, M. M., ... & Constable, R. T. (2015). Functional connectome fingerprinting: identifying individuals using patterns of brain connectivity. *Nature Neuroscience, 18*(11), 1664–1671.

Fischer, R., & Plessow, F. (2015). Efficient multitasking: parallel vs. serial processing of multiple tasks. *Frontiers in Psychology*. https://doi.org/10.3389/fpsyg.2015.01366.

Frank, M. J., & Claus, E. D. (2006). Anatomy of decision: striato-orbitofronal interactions in reinforcement learning, decision making and reversal. *Psychological Review, 113*(2), 300–326.

Franke, A. G., & Lieb, K. (2010). Pharmakologisches Neuroenhancement und „Hirndoping". *Bundesgesundheitsblatt - Gesundheitsforschung - Gesundheitsschutz, 53*(8), 853–860.

Friedman, T. L. (2006). *Die Welt ist flach. Eine kurze Geschichte des 21. Jahrhunderts*. Frankfurt: Suhrkamp Verlag.

Gerrig, R. J. & Zimbardo, P. G. (2007). *Psychology* (18th updated edition). Munich: Pearson Education Germany GmbH.

Gielas, A. (2013). Hat Langeweile einen Sinn? *Psychology Today, 40*(10), 40–44.

Gobet, F., & Simon, H. A. (1998). Expert chess memory: revisiting the chunking hypothesis. *Memory, 6*(3), 225–255.

Godden, D. R., & Baddeley, A. D. (1975). *Context dependent memory in two natural environments: on land and underwater*. Stirling: University of Stirling. Department of Psychology. Revised manuscript.

Goodwin, D. W., Crane, J. B., & Guze, S. B. (1969). Phenomenological aspects of the alcoholic "blackout". *The British Journal of Psychiatry, 115*(526), 1033–1038.

Goodwin, D. W., Powell, B., Bremer, D., Hoine, H., & Stern, J. (1969). Alcohol and recall: State-dependent effects in man. *Science, 163*(3873), 1358–1360.

Görz, G., Rollinger, C.-R., & Schneeberger, J. (2003). *Handbuch der künstlichen Ingelligenz.* (4th corrected edition). Munich Vienna: Oldenbuurg Wissenschaftsverlag.

Graf, O. (1932). Zur Frage der Konstanz bei Intelligenzprüfungen nach verschiedenen Prüfungssystemen. *Psychotechnische Zeitschrift, 7*, 84–88.

Gramann, K., & Schandry, R. (2009). *Psychophysiologie – körperliche Indikatoren psychischen Geschehens.* Weinheim-Basel: Beltz Publishers.

Green, M. F., Lee, J., Wynn, J. K., & Mathis, K. I. (2011). Visual masking in schizophrenia: overview and theoretical implications. *Schizophrenia Bulletin, 37*(4), 700–708.

Greiner, K. (2007). Ist die psychoanalytische Neurophorie begründet? Zur epistemologischen Achillesferse der Neuro-Psychoanalyse. *Psychotherapy Forum, 15*, 134–140.

Greiner, K. (2012). *Standardisierter Therapieschulendialog (TSD). Therapieschulen-interdisziplinäre Grundlagenforschung an der Sigmund Freud Privatuniversität.* Vienna, Sigmund-Freud-Privatuniversitätsverlag.

Greiner, K., & Jandl, M. J. (2015). *Bizarrosophie – Radikalkreatives Forschen im Dienste der akademischen Psychotherapie.* (1st ed.). Nordhausen: Traugott Bautz.

Gross, G. G., Straub, C., Perez-Sanchez, J., Dempsey, W. P., Junge, J. A., Roberts, R. W., ... & Sabatini, B. L. (2016). An E3-ligase-based method for ablating inhibitory synapses. *Nature Methods, 13*, 673–678.

Grossman, P., Niemann, L., Schmidt, S., & Walach, H. (2004). Mindfulness-based stress reduction and health benefits: A meta-analysis. *Journal of Psychosomatic Research, 57*, 35–43.

Guttmann, G. (2000a). Die Messbarkeit des Erlebens und der Irrtum mit dem Ich. In F. Stadler (Ed.), *Elemente moderner Wissenschaftstheorie.* (pp. 73–91). Vienna/New York: Springer.

Guttmann, G. (2000b). Ludwig von Bertalanffys Systemtheorie und die „Wirklichkeit" psychologischer Konzepte. In K. Edlinger, W. Feigl, & G. Fleck (Eds.), *Systemtheoretische Perspektiven.* Frankfurt am Main: Peter Lang. Manuscript of the author.

Guttmann, G. (2013). Von der Erregungssumme zum kortikalen Gleichspannungspotential. Ein Rückblick nach 100 Jahren. *SFU Forschungsbulletin, 1*(1).

Guttmann, G. (2016). Erfinden statt entdecken. In N. Holzenthal (Ed.), *Constructing reality/ Realität konstruierend* (pp. 25–32). Frankfurt am Main: Peter Lang Publishers.

Hamm, S., Junglas, K., & Bredenkamp, J. (2004). Die Zentrale Exekutive als präartikulatorische Kontrollinstanz. *Zeitschrift für Psychologie, 212*(2), 66–75.

Hammond, D. C. (2007). What is neurofeedback? *Journal of Neurotherapy, 10*(4), 25–36.

Hasselhorn, M., & Pit, D. (2007). Was hat das Arbeitsgedächtnis mit dem Erwerb des Lesens, Schreibens und Rechnens zu tun? Wien. In K. Rosenberger & M. Ochoko-Stastny (Eds.), *Mit Sprache wachsen* (pp. 43–59). Vienna: Österreichische Gesellschaft für Sprachheilpädagogik.

Häusser, M., Raman, I. M., Otis, T., Smith, S. L., Nelson, A., Du Lac, S., ... & Yarom, Y. (2004). The beat goes on: spontaneous firing in mammalian neuronal microcircuits. *Journal of Neuroscience, 24*(42), 9215–9219.

Haxby, J. V., Gobbini, M. I., Furey, M. L., Ishai, A., Schouten, J. L., & Pietrini, P. (2001). Distributed and overlapping representations of faces and objects in ventral temporal cortex. *Science, 293*(5539), 2425–2430.

Hebb, D. (2009 [1949]). *The Organization of behavior: A neuropsychological theory*. Mahwah: Lawrence Erlbaum Associates Inc..

Hejl, P. M. (1992). Konstruktion der sozialen Konstruktion – Grundlinien einer konstruktivistischen Sozialtheorie. In H. Gumin & H. Meier (Eds.), *Einführung in den Konstruktivismus* (15th ed., pp. 109–146). Munich: Piper Publishers.

Hills, T. T., Todd, P. M., & Goldstone, R. L. (2010). The central executive as a search process: Priming exploration and exploitation across domains. *Journal of Experimental Psychology: General, 139*(4), 590–609.

Hirsch, J., Kim, K. H., Relkin, N. R., & Lee, K. M. (1997). Distinct cortical areas associated with native and second languages. *Nature, 388*, 171–174.

Ito, T. A., Chiao, K. W., Devine, P. G., Lorig, T. S., & Cacioppo, J. T. (2006). The influence of facial feedback on race bias. *Psychological Science, 17*(3), 256–261.

Jais, A., Solas, M., Backes, H., Chaurasia, B., Kleinridders, A., Theurich, S., ... & Alber, J. (2016). Myeloid-cell-derived VEGF maintains brain glucose uptake and limits cognitive impairment in obesity. *Cell, 165*(4), 882–895.

Jans, M. (2004). Empirische Effekte organisationsdemografischer Diversität in Organisationen – Ergebnisse und Erkenntnisse einer Metaanalyse. Essener Beiträge zur Personalforschung Nr.3. *In soFid Organisations- und Verwaltungsforschung 2006/2*. (pp. 9–43). GESIS-Leibniz-Institute for Social Sciences.

Ji, D., & Wilson, M. A. (2007). Coordinated memory replay in the visual cortex and hippocampus during sleep. *Nature Neuroscience, 10*(1), 100–107.

Kahle, W. & Frotscher, M. (2005). *Nervensysteme und Sinnesorgane*. In: Taschenatlas Anatomie. 3rd band (9th revised edition). Stuttgart. Thieme Publishers.

Kempermann, G., Chesler, E. J., Lu, L., Williams, R. W., & Gage, F. H. (2006). Natural variation and genetic covariance in adult hippocampal neurogenesis. *Proceedings of the National Academy of Sciences, 103*(3), 780–785.

Kennedy, D. (2004). Neuroscience and neuroethics. *Science, 306*(5695), 373.

Kim, K. H. (2005). Can only intelligent people be creative? A meta-analysis. *Journal of Advanced Academics, 16*(2–3), 57–66.

Kimbrel, N. A., Morissette, S. B., Meyer, E. C., Chrestman, R., Jamroz, R., Silvia, P. J., ... & Young, K. A. (2015). Effect of the 5-HTTLPR polymorphism on posttraumatic stress disorder, depression, anxiety, and quality of life among Iraq and Afghanistan veterans. *Anxiety, Stress and Coping, 28*(4), 456–466.

Kitamura, T., Saitoh, Y., Takashima, N., Murayama, A., Niibori, Y., Ageta, H., ... & Inokuchi, K. (2009). Adult neurogenesis modulates the hippocampus-dependent period of associative fear memory. *Cell, 139*(4), 814–827.

Koechlin, E., & Hyafil, A. (2007). Anterior prefrontal function and the limits of human decision-making. *Science, 318*(5850), 594–598.

Krajewski, K., Schneider, W., & Nieding, G. (2008). Zur Bedeutung von Arbeitsgedächtnis, Intelligenz, phonologischer Bewusstheit und früher Mengen-Zahlen-Kompetenz beim Übergang Kindergarten in Grundschule. *Psychologie in Erziehung und Unterricht, 55*(2), 100–113.

Labov, W. (1973): The boundaries of words and their meanings. In: Bailey, C.-J.N. & Shuy, R.W. (Hg.): New ways of analyzing variation in English. Washington, DC: Georgetown University Press. 340-373.

Langer, E., Blank, A., & Chanowitz, B. (1978). The mindlessness of ostensibly thoughtful action: The role of "placebic" information in interpersonal interaction. *Journal of Personality and Social Psychology, 36*(6), 635–642.

Lee, J., Duan, W., & Mattson, M. P. (2002). Evidence that brain-derived neurotrophic factor is required for basal neurogenesis and mediates, in part, the enhancement of neurogenesis by dietary restriction in the hippocampus of adult mice. *Journal of Neurochemistry, 82*(6), 1367–1375.

Lerner, Y., Hendler, T., Ben-Bashat, D., Harel, M., & Malach, R. (2001). A hierarchical axis of object processing stages in the human visual cortex. *Cerebral Cortex, 11*(4), 287–297.

Lewis, M. L., & Frank, M. C. (2016). The length of words reflects their conceptual complexity. *Cognition, 153*, 182–195.

Libet, B. (1985). Unconscious cerebral initiative and the role of conscious will in voluntary action. *The Behavioral and Brain Sciences, 8*(4), 529–566.

Libet, B., Gleason, C. A., Wright, E. W., & Pearl, D. K. (1983). Time of conscious intention to act in realisation to onset of cerebral activity (readiness potential): the unconscious initiation of a freely voluntary act. *Brain, 106*, 623–642.

Luck, S. J., Hillyard, S. A., Mouloua, M., Woldorff, M. G., Clark, V. P., & Hawkins, H. L. (1994). Effects of spatial cuing on luminance detectability: psychophysical and electrophysiological evidence for early selection. *Journal of Experimental Psychology: Human Perception and Performance, 20*(4), 887–904.

Macedonia, M., & Höhl, S. (2012). *Gehirn für Einsteiger* (2nd ed.). Linz/Heidelberg: Self-published.

Macedonia, M., & Höhl, S. (2013). *Gehirn für Fortgeschrittene.* Linz: SchEz -School and Education Centre Linz.

Maeda, F., Keenan, J. P., Tormos, J. M., Topka, H., & Pascual-Leone, A. (2000). Interindividual variability of the modulatory effects of repetitive transcranial magnetic stimulation on cortical excitability. *Experimental Brain Research, 133*(4), 425–430.

Maguire, E. A., Woollett, K., & Spiers, H. J. (2006). London taxi drivers and bus drivers: A structural MRI and neuropsychological analysis. *Hippocampus, 16*, 1091–1101.

Majumdar, M., Grossman, P., Dietz-Waschkowski, B., Kersig, S., & Walach, H. (2002). Does mindfulness meditation contribute to health? Outcome evaluation of a German sample. *The Journal of Alternative and Complementary Medicine, 6*, 719–730.

Marchand, W. R. (2014). Neural mechanisms of mindfulness and meditation: evidence from neuroimaging studies. *World Journal of Radiology, 6*(7), 471–479.

Maturana, H. R., & Varela, F. J. (2015). *Der Baum der Erkenntnis* (6th ed.). Frankfurt am Main: Fischer Paperback Publisher.

Mednick, S. (1962). The associative basis of the creative process. *Psychological Review, 69*(3), 220–232.

Mednick, S. A., & Mednick, M. T. (1967). *Examiner's manual, remote associates test.* Boston: Houghton Mifflin.

Metha, M., & Schmauss, C. (2011). Strain-specific cognitive deficits in adult mice exposed early life stress. *Behavioral Neuroscience, 125*(1), 29–36.

Mogge-Grotjahn, H. (2011). *Soziologie: eine Einführung für soziale Berufe.* Freiburg im Breisgau: Lambertus Publishing House.

Moore, T., & Armstrong, K. M. (2003). Selective gating of visual signals by microstimulation of frontal cortex. *Nature, 421*, 370–373.

Morey, R. A., Gold, A. L., LaBar, K. S., Beall, S. K., Brown, V. M., Haswell, C. C., ... & McCarthy, G. (2012). Amygdala volume changes in posttraumatic stress disorder in a large case-controlled veterans group. *Archives of General Psychiatry, 69*(11), 1169–1178.

Muir, D. W., Blasdel, G. G., Mitchell, D. E., & Pettigrew, J. D. (1977). A physiological and behavioural study in cats of the effect of early visual experience with contours of a single orientation. *Journal of Physiology, 265*, 615–636.

Navon, D. (1977). Forest before trees: the precedence of global features in visual perception. *Cognitive Psychology, 9*, 353–383.

O'Connor, B. P., & Molly, K. (1991). A test of the intellectual cycle of the popular biorhythm theory. *The Journal of Psychology: Interdisciplinary and Applied, 125*(3), 291–299.

Pavlov, I. P. (2010 [1927]). Conditioned reflexes: An investigation of the physiological activity of the cerebral cortex. *Annals of Neurosciences, 17*(3), 136–141.

Pease, A., & Pease, B. (2004). *The definitive book of body language.* New York: Bantam Dell.

Perani, D., Paulesu, E., Galles, N. S., Dupoux, E., Dehaene, S., Bettinardi, V., ... & Mehler, J. (1998). The bilingual brain. Proficiency and age of acquisition of the second language. *Brain: a journal of neurology, 121*, 1841–1852.

Pettigrew, J. D. (1974). The effect of visual experience on the development of stimulus specificity by kitten cortical neurones. *Journal of Physiology, 237*, 49–74.

Pierucci-Lagha, A., Covault, J., Feinn, R., Nellissery, M., Hernandez-Avila, C., Oncken, C., ... & Kranzler, H. R. (2005). GABRA2 alleles moderate the subjective effects of alcohol, which are attenuated by finasteride. *Neuropsychopharmacology, 30*, 1193–1203.

Plessow, F., Kiesel, A., & Kirschbaum, C. (2012). The stressed prefrontal cortex and goal-directed behaviour: acute psychosocial stress impairs the flexible implementation of task goals. *Experimental Brain Research, 216*(3), 397–408.

Post, A., Müller, M. B., Engelmann, M., & Keck, M. E. (1999). Repetitive transcranial magnetic stimulation in rats: evidence for a neuroprotective effect in vitro and in vivo. *European Journal of Neuroscience, 11*(9), 3247–3254.

Prinz, M., Erny, D., de Angelis, A. L. H., Jaitin, D., Wieghofer, P., Staszewski, O., ... & Schwierzeck, V. (2015). Host microbiota constantly control maturation and function of microglia in the CNS. *Nature Neuroscience, 18*, 965–977.

Psychotherapiegesetz. BGBl. Nr. 361/1990 as amended on. Wien.

Rensink, R. A., O'Regan, J. K., & Clark, J. J. (1997). To see or not to see: the need for attention to perceive changes in scenes. *Psychological Science, 8*(5), 368–373.

Ricken, G., & Fritz, A. (2006). Arbeitsgedächtnisleistung bei unterschiedlich guten Rechnern im Kindergartenalter. *Psychologie in Erziehung und Unterricht, 53*(4), 263–274.

Rieken, B. (2010). Das Analogiedenken und seine Bedeutung für Medizin und Psychotherapie. Potsdam. AGEM (Arbeitsgemeinschaft Ethnomedizin e.V.). *Curare, 33*, 192–199.

Ritter, S. M., Damian, R. I., Simonton, D. K., van Baaren, R. B., Strick, M., Derks, J., & Dijksterhuis, A. (2012). Diversifying experiences enhance cognitive flexibility. *Journal of Experimental Social Psychology, 48*, 961–964.

Rösler, F. (2011). *Psychophysiologie der Kognition. Eine Einführung in die Kognitive Neurowissenschaft*. Heidelberg: Spectrum Academic Publishing House.

Roth, G., & Strüber, N. (2014). *Wie das Gehirn die Seele macht* (5th ed.). Stuttgart: Klett-Cotta Publisher.

Sanders, L. (2015). Nets full of holes catch long-term memories. *Science News, 188*(10), 8.

Schandry, R. (2011). *Biologische Psychologie* (3rd completely revised edition). Weinheim-Basel: Beltz Publishers.

Schenk-Danzinger, L. (2006). *Entwicklungspsychologie*. Vienna: G&G Publishing Company. (2nd edition in complete revision by Karl Rieder. Reprint 2011).

Schmidt, S. J. (1992). Vom Text zum Literatursystem – Skizze einer konstruktivistischen (empirischen) Literaturwissenschaft. In H. Gumin & H. Meier (Eds.), *Einführung in den Konstruktivismus* (15th ed., pp. 147–166). Munich: Piper Publishers.

Shansky, R. M., & Lipps, J. (2013). Stress-induced cognitive dysfunction: hormone-neurotransmitter interactions in the prefrontal cortex. *Frontiers in Human Neuroscience, 7*, 278–283.

Shors, T. J. (2009). Saving new brain cells. *Scientific American, 300*(3), 47–54.

Siegler, R. S. (2000). Unconscious insights. *Current Directions in Psychological Science, 9*(3), 79–83.

Silbernagl, S. & Despopoulos, A. (2003). *Taschenatlas der Physiologie* (6th corrected edition). Stuttgart: Thieme Publishers.

Silver, I. A., Deas, J., & Erecinska, M. (1997). Ion homeostasis in brain cells: differences in intercellular ion responses to energy limitation between cultured neurons and glial cells. *Neuroscience, 78*(2), 589–601.

Simons, D. J., & Chabris, C. F. (1999). Gorillas in our midst: sustained inattentional blindness for dynamic events. *Perception, 28*(9), 1059–1074.

Singer, J. A., & Salovey, P. (1988). Mood and memory: evaluating the network theory of affect. *Clinical Psychological Review, 8*(2), 211–251.

Sligh, A. C., Conners, F. C., & Roskos-Ewoldsen, B. (2005). Relation of creativity to fluid and crystallized intelligence. *The Journal of Creative Behavior, 39*(2), 123–136.

Sliwinski, M. J., Smyth, J. M., Hofer, S. M., & Stawski, R. S. (2006). Intraindividual coupling of daily stress and cognition. *Psychology and Aging, 21*(3), 545–557.
Smith, I. L. (1971). IQ, creativity, and achievement: interaction and threshold. *Multivariate Behavioral Research, 6*(1), 51–62.
Smith, S. M., & Vela, E. (2001). Environmental context-dependent memory: a review and meta-analysis. *Psychonomic Bulletin & Review, 8*(2), 202–220.
Spitzer, M. (2009). *Lernen – Gehirnforschung und die Schule des Lebens*. Heidelbert: Spectrum Academic Publishing House.
Spitzer, M. (2012). *Digitale Demenz*. Munich: Droemer Publishers.
Stankiewicz, A. M., Swiergiel, A. H., & Lisowski, P. (2013). Epigenetics of stress adaptations in the brain. *Brain Research Bulletin, 98*, 76–92.
Sterman, M. B., & Egner, T. (2006). Foundation and practice of neurofeedback for the treatment of epilepsy. *Applied Psychophysiology and Biofeedback, 31*(1), 21–35.
Strack, F., Martin, L. L., & Stepper, S. (1988). Inhibiting and facilitating conditions of the human smile: a nonobtrusive test of the facial feedback hypothesis. *Journal of Personality and Social Psychology, 54*(5), 768–777.
Tang, Y. Y. (2011). Mechanism of integrative body-mind training. *Neuroscience Bulletin, 27*(6), 383–388.
Tang, Y. Y., Ma, Y., Fan, Y., Feng, H., Wang, J., Feng, S., ... & Zhang, Y. (2009). Central and autonomic nervous system interaction is altered by short-term meditation. *Proceedings of the National Academy of Sciences, 106*(22), 8865–8870.
Tang, Y. Y., Lu, Q., Geng, X., Stein, E. A., Yang, Y., & Posner, M. I. (2010). Short-term meditation induces white matter changes in the anterior cingulate. *Proceedings of the National Academy of Sciences, 107*(35), 15649–15652.
Taverniers, J., Taylor, M. K., & Smeets, T. (2013). Delayed memory effects after intense stress in Special Forces candidates: exploring path processes between cortisol secretion and memory recall. *Stress, 16*(3), 311–320.
Thatcher, R. W., North, D. M., & Biver, C. J. (2008). Intelligence and EEG phase reset: a two compartmental model of phase shift and lock. *Neuroimage, 42*, 1639–1653.
Tsai, H. H., Niu, J., Munji, R., Davalos, D., Chang, J., Zhang, H., ... & Fancy, S. P. (2016). Oligodendrocyte precursors migrate along vasculature in the developing nervous system. *Science, 351*(6271), 379–383.
Underwood, E. (2014). Starting young – decades old IQ test records from Scotish children have opened a unique window on how the brain ages. *Science, 346*(6209), 568–571.
von der Malsburg, C. (1973). Self-organization of orientation sensitive cells in the striate cortex. *Cybernetics, 14*(2), 85–100.
von Foerster, H. (1992). Entdecken oder Erfinden – Wie lässt sich Verstehen verstehen?. In H. Gumin & H. Meier (Eds.), *Einführung in den Konstruktivismus*. (15th ed., pp. 41–88). Munich: Piper Publishers.
von Glasersfeld, E. (1992). *Konstruktion der Wirklichkeit und des Begriffs der Objektivität*. In H. Gumin & H. Meier (Eds.), *Einführung in den Konstruktivismus*. (15th ed., pp. 9–40). Munich: Piper Publishers.
Vock, M., & Hasselhorn, M. (2010). Diagnostik der Funktionstüchtigkeit des Arbeitsgedächtnisses bei Hochbegabten. In F. Preckel, W. Schneider, & H. Holling (Eds.), *Diagnostik von Hochbegabung* (pp. 119–139). Göttingen: Hogrefe.
Wagner, M. (2007). *Wir sehen mit den Augen des Kollektivs!?* Bad Heilbrunn: Julius Klinkhardt.
Wallner, F. (2002). *Die Verwandlung der Wissenschaft*. Jandl, M.J. (Hg.). Hamburg: Dr. Kovač.
Wallner, F. G., & Jandl, M. J. (2006). The importance of constructive realism for the indigenous psychologies approach. In U. Kim, K.-S. Yang, & K. K. Hwang (Eds.), *Indigenous and cultural psychology* (pp. 49–72). New York: Springer.
Wan, C. Y., & Schlaug, G. (2010). Music making as a tool for promoting brain plasticity across the life span. *The Neuroscientist, 16*(5), 566–577.

Wassermann, E. M. (1998). Risk and safety of repetitive transcranial magnetic stimulation: report and suggested guidelines from the International Workshop on the Safety of Repetitive Transcranial Magnetic Stimulation. *Electroencephalography and Clinical Neurophysiology, 108*(1998), 1–16.

Watson, D., Clark, L. A., & Tellegen, A. (1988). Development and validation of brief measures of positive and negative affect: the PANAS Scales. *Journal of Personality and Social Psychology, 54*(6), 1063–1070.

Weiller, C., Jüptner, M., Fellows, S., Rijntjes, M., Leonhardt, G., Kiebel, S., ... & Thilmann, A. F. (1996). Brain representation of active and passive movements. *Neuroimage, 4*, 105–110.

Wolf, S. A., Möhle, L., Mattei, D., Heimesaat, M. M., Bereswill, S., Fischer, A., ... & Alutis, M. (2016). Ly6Chi monocytes provide a link between antibiotic-induced changes in gut microbiota and adult hippocampal neurogenesis. *Cell Reports, 15*(9), 1945–1956.

Xie, P., Kranzler, H. R., Poling, J., Stein, M. B., Anton, R. F., Brady, K., ... & Gelernter, J. (2009). Interactive effect of stressful life events and the serotonin transporter 5-HTTLPR genotype on posttraumatic stress disorder diagnosis in 2 independent populations. *Archives of General Psychiatry, 66*(11), 1201–1209.

Ylönen, H., Lyytinen, H., Leino, T., Leppäluoto, J., & Kuronen, P. (1997). Heart rate responses to real and simulated BA Hawk MK 51 flight. *Aviation, Space and Environmental Medicine, 68*(7), 601–605.

Yu, S. P., Wang, X. Q., Xiao, A. Y., Sheline, C., Hyrc, K., Yang, A., ... & Goldberg, M. P. (2003). Apoptotic insults impair Na+, K+-ATPase activity as a mechanism of neuronal death mediated by concurrent ATP deficiency and oxidant stress. *Journal of Cell Science, 116*(10), 2099–2110.

Zeilinger, A. (2005). *Einsteins Schleier – Die neue Welt der Quantenphysik*. Munich: Wilhelm Goldmann Publishers.

Online Resources

Bartol Jr, T. M., Bromer, C., Kinney, J., Chirillo, M. A., Bourne, J. N., Harris, K. M., & Sejnowski, T. J. (2015). Nanoconnectomic upper bound on the variability of synaptic plasticity. *eLife, 4*, e10778.

Brodbeck, K. H. (2006). Neue Trends in der Kreativitätsforschung. In *Themenschwerpunkt Kreativität, Kunst und Humor.* BÖP. Psychologie in Österreich, pp. 246-253. http://www.khbrodbeck.homepage.t-online.de/trends.pdf. Accessed 2 June 2016.

Friedman, T. L. (2006): *Die Welt ist flach. Eine kurze Geschichte des 21. Jahrhunderts.* Frankfurt: Suhrkamp Verlag.

Glasser, M. F., Coalson, T. S., Robinson, E. C., Hacker, C. D., Harwell, J., Yacoub, E., ... & Smith, S. M. (2016). A multi-modal parcellation of human cerebral cortex. *Nature, 536* (7615), 171-178. https://doi.org/10.1038/nature18933.

Guttmann, G. (2013): Von der Erregungssumme zum kortikalen Gleichspannungspotential. Ein Rückblick nach 100 Jahren. *SFU Forschungsbulletin 1*(1).

Haxby, J. V., Gobbini, M. I., Furey, M. L., Ishai, A., Schouten, J. L., & Pietrini, P. (2001). Distributed and overlapping representations of faces and objects in ventral temporal cortex. *Science, 293*(5539), 2425–2430.

Hebb, D. (2009 [1949]). *The Organization of behavior: A neuropsychological theory.* Mahwah: Lawrence Erlbaum Associates Inc.

Herff, C., Heger, D., De Pesters, A., Telaar, D., Brunner, P., Schalk, G., & Schultz, T. (2015). Brain-to-text: decoding spoken phrases from phone representations in the brain. *Frontiers in Neuroscience, 9*, 217. https://doi.org/10.3389/fnins.2015.00217.

Kempermann, G., Fabel, K., Ehninger, D., Babu, H., Leal-Galicia, P., Garthe, A., & Wolf, S. (2010). Why and how physical activity promotes experience-induced brain plasticity. *Frontiers in Neuroscience, 4*, 189. https://doi.org/10.3389/fnins.2010.00189.

Koechlin, E., & Hyafil, A. (2007). Anterior prefrontal function and the limits of human decision-making. *Science, 318*(5850), 594–598.

Kreilhuber, A. (2014). *Körperliche Auswirkungen von Alkohol*. http://www.netdoktor.at/krankheit/koerperliche-auswirkungen-von-alkohol-6451633. Accessed 25 June 2016.

Labov, W. (1973): The boundaries of words and their meanings. In: Bailey, C.-J.N. & Shuy, R.W. (Hg.): New ways of analyzing variation in English. Washington, DC: Georgetown University Press. 340–373.

Limb, C. J., & Braun, A. R. (2008). Neural substrates of spontaneous musical performance: an fMRI study of Jazz improvisation. *PLoS One, 3*(2), e1679.

Liu, S., Chow, H. M., Xu, Y., Erkkinen, M. G., Swett, K. E., Eagle, M. W., ... & Braun, A. R. (2012). Neural correlates of lyrical improvisation: an fMRI study of freestyle rap. *Scientific Reports, 2*, 834.

Manning, J. R. (2016). A neural signature of contextually mediated intentional forgetting. *Psychonomic Bulletin & Review, 2016*, 1–31.

Mars1. (2013). *Skylab: America's first space station*. http://ablogaboutnothinginparticular.com/?=1042. Accessed 1 Aug 2016.

Pavlov, I. P. (2010 [1927]). Conditioned reflexes: An investigation of the physiological activity of the cerebral cortex. *Annals of Neurosciences, 17*(3), 136–141.

Piore, A. (2013). When brain damage unlocks the genius within. *Popular Science, 2013*. https://www.popsci.com/science/article/2013-02/when-brain-damage-unlocks-genius-within/ Accessed 21 Nov 2020.

Rash, B. G., Ackman, J. B., & Rakic, P. (2016). Bidirectional radial Ca^{2+} activity regulates neurogenesis and migration during early cortical column formation. *Science Advanced, 2*, 1–10.

Unknown (2012). *Living in space*. Homepage of the European Space Agency. http://www.esa.int/Our_Activities/Human_Spaceflight/Astronauts/Living_in_space. 1 Aug 2016.

Underwood, E. (2015). *Mysterious 'holes' in neuron net may help store long-term memories*. Washington, DC: American Association for the Advancement of Science. http://www.sciencemag.org/news/2015/10/mysterious-holes-neuron-net-may-help-store-long-term-memories. Accessed 27 Dec 2015.

Vock, M., & Hasselhorn, M. (2010). Diagnostik der Funktionstüchtigkeit des Arbeitsgedächtnisses bei Hochbegabten. In F. Preckel, W. Schneider, & H. Holling (Eds.), *Diagnostik von Hochbegabung* (pp. 119–139). Göttingen: Hogrefe.

Zsynth (2006). *Figure lateral inhibition*. From the German Wikipedia, CC BY-SA 3.0 [vom 27 07 16] https://commons.wikimedia.org/w/index.php?curid=1645056.

Other

Gramann, K. (2002). *Arbeitsgedächtnis und elektrokortikale Aktivität – Eine experimentelle Untersuchung zur Differenzierung von Subsystemen des Arbeitsgedächtnisses*. Dissertation. Aachen: Faculty of Philosophy, Rheinländisch-Westfälische Technische Hochschule Aachen.

Greiner, K. (2013). *Konstruktiv-realistische Psychotherapiewissenschaft*. Vienna/Paris: Teaching Script of the SFU Sigmund-Freud-Privatuniversität Wien/Paris.

Guttmann, G. (1995). *Psychophysiology of hypnosis*. Manuscript of the author.

Liessmann, K. P. (2016). Kunst als Manifest. Plädoyer Salzburger Festspiele. In *Kleine Zeitung of 29 July 2016* (pp. 2–5). Graz: Kleine Zeitung GmbH & Co KG.

Nicolosi, L., Eggert, J., Riggenbach, S., Khan-Mayberry, N., Wilson, R., & Hensley, J. (2015). *NASA's effort to manage health and human performance risks for space exploration*. Washington, DC: NASA Office of Inspector General-Office of Audits. Report No. IG16-003.

Psychotherapiegesetz. BGBl. Nr. 361/1990 as amended on. Wien.

Seifert, G. (2009). *Langzeitpotenzierung in der CA1-Region – Verschiedene Modulatoren.* Inaugural dissertation. Freiburg: Medical Faculty of the Albert-Ludwigs-University.

Stephenson, T. (2014). *Deep impact.* Workshop on ZRD: Citation/Reception/Discourse. Vienna: SFU. 30/31 January 2014.

Täuber, M. (2016). *Course to become a qualified neuromental trainer. Seminar units 1–4 from February to June 2016.* Vienna: Institute for Mental Success Strategies.

Wassermann, E. M. (1998): Risk and safety of repetitive transcranial magnetic stimulation: report and suggested guidelines from the International Workshop on the Safety of Repetitive Transcranial Magnetic Stimulation. *Electroencephalography and Clinical Neurophysiology 108* (1998), 1–16.

Wiesner, C. D. (2004). *Was ist zentral an der zentralen Exekutive? Neuronale Korrelate des vorhersagbaren versus unvorhersagbaren kognitiven Verzweigens in verbal-phonologischen versus figural-räumlichen Arbeitsgedächtnisaufgaben. Eine Studie mit funktioneller Magnetresonanztomographie.* Dissertation. Kiel: Christian Albrecht University.

Index

K. Testor, *Cognitive Theoretical Foundations of Creativity*,
https://doi.org/10.1007/978-3-030-61736-3

D

The manufacturer's authorised representative in the EU is Springer Nature Customer Service Centre GmbH, Europaplatz 3, 69115 Heidelberg, Germany. If you have any concerns regarding our products, please contact ProductSafety@springernature.com

Printed and bound by CPI Group (UK) Ltd, Croydon, CR0 4YY

30/09/2026

02225320-0007